Eddie's Cheezy Notes on Organic Chemistry

An Illustrated guide on Reactions and Reagents for Both Non-majors and Educators

Edward Antonio

Cover design and artwork:
Pauline Antonio
Editor:
Christina Collison, Ph.D.
Consulting editor:
Claudiu Dumitrescu, Ph.D.
Proof readers/reviewers:
Elizabeth Dillon
Haeja Kessler
Contributions by:
Joseph P. Hornak, Ph.D.

Acknowledgments

I would like to thank all my Chemistry professors who have generously imparted their knowledge and consequently have inspired me to write this book. I especially owe a great deal of thanks to my Organic Chemistry professor, Dr. Brian Edelbach, whose projects encouraged me to write what turned out to be my best cheesy notes. Special thanks also goes to Dr. Claudiu Dumitrescu for his attention to details and editing support. To my relatives, friends, and of course my family – Pauline, Gabrielle, Giselle, and Gweneth, your support means everything. Lastly, I thank my Lord for helping me and giving me the wisdom for this project

Preface

Before I take you on this journey, I want to set the stage by imparting to you a small measure of my philosophy of learning. I firmly believe that the key to your success lies in your alliance with your instructors. Whether you agree with their teaching style or not, they are your guide through the quagmire of formulas, theoretical concepts, and laboratory experiments that often make up these courses. They have the understanding and experience to know what you need to do in order to be successful. Once you come to this realization and start with the attitude that says "I will reach out for help when I need it", you're blood pressure will remain under control, and that means less stress and better concentration.

As I'm sure you know by now, organic chemistry has many areas of study that include resonance structures, arrow pushing, delocalization, acids & bases, reactions, and much more. This book was structured to guide you through the study of organic reactions and reagents. Many steps may be required to reach the end goal of a desired reaction outcome. To get there, one must have the basic understanding and also learn to master one reaction at a time.

"Eddie's Cheezy Notes on Organic Chemistry" simplifies and shortens the seemingly overwhelming concepts. It is meant to replace the wordy supplemental material that often is just as confusing as the main text. You'll have enough confusing material to digest in your text book and class handouts. In sum, this book is designed to be a guide to help you comprehend the material with simple explanations.

I clearly remember when I first took Organic Chemistry. Challenging would not even begin to describe my experience. However, by the second semester I began to see the light. I found myself under the tutelage of Dr. Edelbach who had us draw equations of reactants and annotate each one with reasons as to why it was important to remember. Although this project was worth only 20 points out of a total of 500 that made up our grade, it proved to be the most valuable. I came to realize that this simple exercise was the catalyst that helped get the equations and key concepts stuck in my noggin. My grades consequently improved as evidenced by my exam and quiz scores that rose to and remained in the 90's throughout the course. I subsequently started sharing my revelation with classmates who quickly came to benefit from this approach.

So here is my suggestion on how to use this book:

This book requires a two-step process. One is your commitment to succeed in your course which means you must first study the concepts your instructor is teaching (i.e. alkanes, ketones, epoxides) along with any assigned handouts or book chapters. Parallel your handouts and book chapter study with a review of relevant material in this book. This should help you make sense of your text book. Second, re-read my cheezy notes a few days before taking an exam. Study the reactions and don't forget to observe the key points. Take care and good fortune in your studies.

Disclaimer:

I know you are smarter than this but I just wanted to say that pepperoni pizza and drinks do not mix well with Diels-Alder, Grignard, or any type of reactions in the lab. Always use gloves and goggles when necessary.

Contents

I. ALCOHOLS

1. Acid-catalyzed hydration of an alkene

Functional Group Transformation: alkene → alcohol

General Scheme:

$$R-CH=CH_2 + H_2O \underset{}{\overset{H_2SO_4}{\rightleftharpoons}} R-CH-CH_2$$
$$\qquad\qquad\qquad\qquad\qquad\qquad OH \quad H$$

Specific Example:

Mechanism:

$$H_2SO_4 + H_2O \rightleftharpoons H_3O^{\oplus} + HSO_4^{-}$$
hydronium ion

hydronium ion

Key points:
- H_2SO_4/H_3O^+, or H_2O with H_2SO_4 as a catalyst
- Catalyst (acid) needed (H_2SO_4, H_3O^+, etc.)
- Carbocation intermediate (rearrangement possible)
- Markovnikov Addition

2. <u>Oxymercuration-demercuration of an alkene</u>

Functional Group Transformation: alkene → alcohol

General Scheme:

$$R - CH = CH_2 \xrightarrow[\text{2) Na BH}_4]{\text{1) Hg (OAc)}_2, H_2O/THF} R - \underset{\underset{OH}{|}}{CH} - CH_3$$

Specific Example:

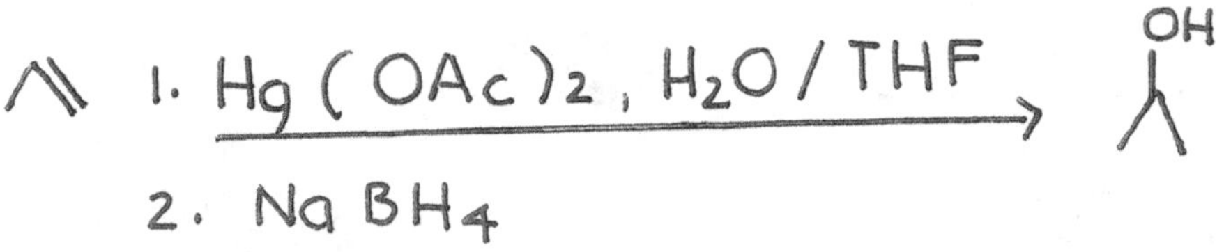

Mechanism:

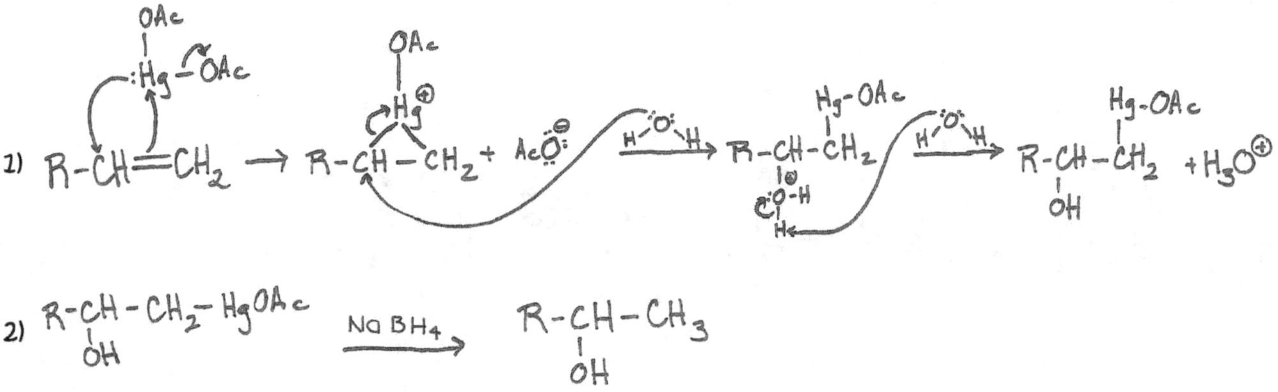

Key points:
- No carbocation intermediate
- Markovnikov addition
- THF used (tetrahydrofuran)

3. ## Hydroboration-oxidation of an alkene

Functional Group Transformation: alkene → alcohol

General Scheme:

$$R-CH=CH_2 \xrightarrow[\text{2. } H_2O_2-OH]{\text{1. } BH_3/THF} R-\underset{\underset{H}{|}}{C}H-\underset{\underset{OH}{|}}{C}H_2$$

Specific Example:

$$\xrightarrow[\text{2. } H_2O_2, {}^-OH]{\text{1. } BH_3/THF}$$

Mechanism:

$$R-CH=CH_2 \xrightarrow{THF} R-\underset{\underset{H}{|}}{C}H-\underset{\underset{BH_2}{|}}{C}H_2 \xrightarrow{H_2O_2,\ {}^-OH} R-CH_2-\underset{\underset{OH}{|}}{C}H_2$$

Key points:
- Anti-Markovnikov Addition
- THF used (tetrahydrofuran)
- Carbocation intermediate not formed
- Syn addition

4. <u>Reaction of an alkyl halide with HO-</u>

Functional Group Transformation: alkyl halide → alcohol

General Scheme:

$$HO:^{\ominus} \quad + \quad R - \ddot{X}: \quad \longrightarrow \quad R - OH \quad + \quad :\ddot{X}:^{\ominus}$$

Specific Example:

$$HO:^{\ominus} \quad + \quad CH_3 - \ddot{B}r: \quad \longrightarrow \quad CH_3 OH \quad + \quad :\ddot{B}r:^{\ominus}$$

Mechanism:

transition state

Key points:
- X = halide (e.g. Br, Cl)
- S_N2 reaction
- Back side attack
- "R" to "S" or from "S" to "R" in most cases (check priorities if chiral)
- No intermediate

5. Reaction of a Grignard reagent with an epoxide

Functional Group Transformation: epoxide → alcohol

General Scheme:

$$R-MgX + R-CH-CH-R \longrightarrow R-CH-CH-R + Mg^{2+} + :\ddot{X}:^{\ominus}$$

(epoxide: O bridging the two CH; product has OH on second carbon and CH_3 on first carbon)

Specific Example:

$$CH_3-MgBr + CH_3-CH-CH-CH_3 \longrightarrow CH_3-CH-CH-CH_3 + Mg^{2+} + :\ddot{Br}:^{\ominus}$$

with CH_3 substituent and OH on product

or

Mechanism:

$MgBr^{+1}$

$$CH_3-MgBr + CH_3-CH-CH-CH_3 \longrightarrow CH_3-CH-CH-CH_3 \longrightarrow CH_3-CH-CH-CH_3$$

Key points:
- No carbocation intermediate
- X = Br or Cl

6. <u>Cleavage of an ether with HI or HBr</u>

Functional Group Transformation: ether → alcohol

General Scheme:

$$R-O-R' + H-X \xrightarrow[HEAT]{\triangle} R-OH + R'-X$$

Specific Examples:

$$\text{EXCESS}$$
$$\diagup^O\diagdown + H-Br \xrightarrow{\triangle} CH_3-OH + CH_3-Br \longrightarrow 2CH_3-Br + H_2O$$

$$\underset{\text{}}{\bigcirc}\diagup^O\diagdown + H-Br \xrightarrow[H_2O]{\triangle} \underset{\text{}}{\bigcirc}\diagup^{OH} + \diagup\diagdown_{Br}$$

Mechanisms:

SN2

$$H_3C-\ddot{O}-CH_3 + H-\overset{\frown}{I} \rightleftharpoons H_3C-\overset{\overset{H}{\oplus}}{O}-CH_3 + \ddot{\underset{..}{I}}{}^{\ominus} \rightarrow CH_3\ddot{O}H + CH_3-I \xrightarrow{H-\overset{\frown}{I}} H_3C-\overset{\oplus}{\underset{}{O}}H_2 \longrightarrow H_3C-I$$
$$\ddot{\underset{..}{I}}{}^{\ominus} \qquad\qquad \overset{+}{H_2O}$$

SN1

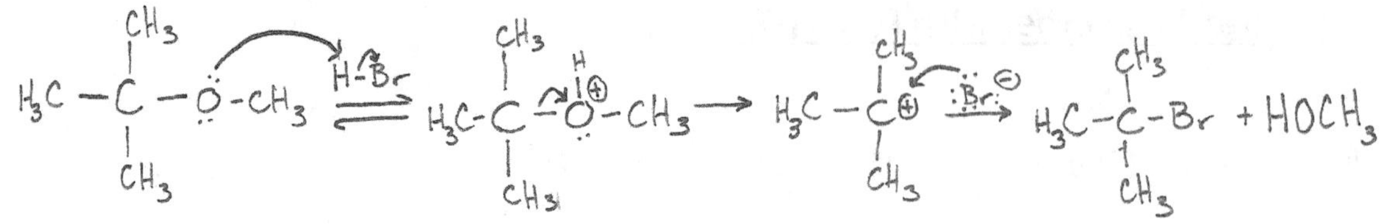

carbocation on 3°

Key points:

- Strong acids used
- Heat required (Δ)
- S_N1 on 3°
- Remember S_N2, "I" or "Br" attacks less hindered base

7. Reduction of an aldehyde to an alcohol

Functional Group Transformation: aldehyde → alcohol

General Scheme:

$$R-\overset{\overset{O}{\|}}{C}H \xrightarrow[\text{(as H}^-\text{source)}]{\text{NaBH}_4 \text{ or LiAlH}_4} R-\overset{\overset{OH}{|}}{\underset{|}{C}}H$$

Specific Example:

$$\overset{O}{\overset{\|}{\diagdown\diagup}}H \xrightarrow{H^{\ominus}} \diagdown\diagup OH$$

Key points:
- Require [H-] from reducing agent
- Reducing agents:
  - Sodium borohydride (NaBH4)
  - Lithium aluminum hydride (LiAlH4)

8. **Nucleophilic substitution reaction of an epoxide**

Functional Group Transformation: epoxide → alcohol

General Scheme:

$$R-CH_2-CH-CH_2 \xrightarrow{H^{\oplus}} R-CH_2-CH-CH_2 \xrightarrow{CH_3OH} R-CH_2-CH-CH_2-OH + R-CH_2-CH-CH_2-OCH_3 + H^{\oplus}$$

(Major Product)   (Minor Product)

2-methoxy-1-propanol + 1-methoxy-2-propanol

Specific Example:

Key points:
- Epoxides **CAN** undergo nucleophilic (nu:) substitution reactions w/out first being protonated because the 3-membered ring causes strain.
- Super Important! Which side does -OCH₃ attack occur? Answer: nucleophile attacks the more substituted ring carbon in acidic environments while under basic environments, the nucleophile attacks the less hindered ring carbon. (This can be on an exam question!) See below:

nu: attack under acidic conditions

nu: attack under basic or neutral conditions

II. Aldehydes

1. Hydroboration-oxidation of a terminal alkyne with Sia_2BH

Functional Group Transformation: alkyne → aldehyde

General Scheme:

$$R-CH_2-C\equiv CH + H_2O \xrightarrow[2.\,H_2O_2,\,^-OH]{1.\,Sia_2BH} R-CH=\overset{\overset{\displaystyle OH}{\displaystyle |}}{CH} \rightleftharpoons R-CH_2\overset{\overset{\displaystyle O}{\displaystyle \|}}{CH}$$

Specific Example:

$$\equiv\!\!\!\!\!\!H + H_2O \xrightarrow[2.\,H_2O_2,\,^-OH]{1.\,Sia_2BH} \diagup\!\!\!\!\diagdown OH \rightleftharpoons \diagup\!\!\!\!\diagdown\!\!\!\diagup^O$$

Mechanism:

Key points:
- Anti-Markovnikov Addition
- Enol formed
- Intermediate forms an enol and ending with an aldehyde
- Syn addition
- 1. Sia_2 BH (Disiamylborane), 2. H_2O_2, -OH

2. Oxidation of a primary alcohol with pyridinium chlorochromate

Functional Group Transformation: alcohol → aldehyde

General Scheme:

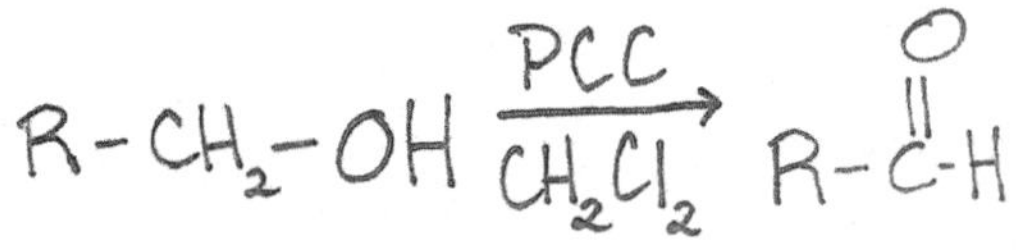

$$R-CH_2-OH \xrightarrow[CH_2Cl_2]{PCC} R-\overset{\overset{O}{\|}}{C}-H$$

Specific Example:

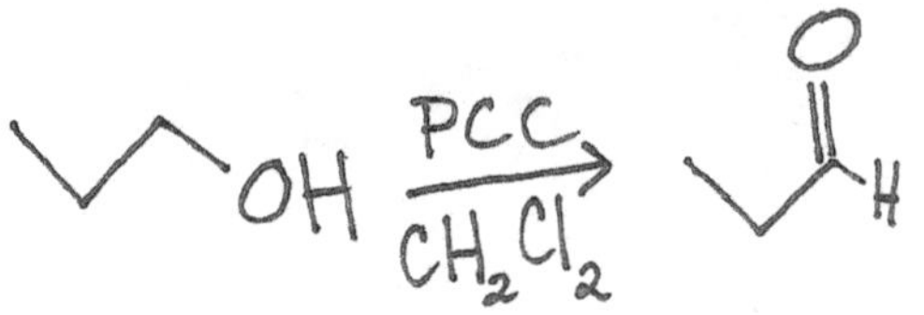

Key points:
- If PCC (pyridinium chlorochromate) not used (H_2CrO_4 in substitution), reaction will not stop at aldehyde but will continue to form carboxylic acids (COOH)
- Primary alcohols gives aldehydes using PCC

3. Reaction using DIABAL-H

Functional Group Transformation: ester → aldehyde
                                    acid chloride → aldehyde

General Schemes:

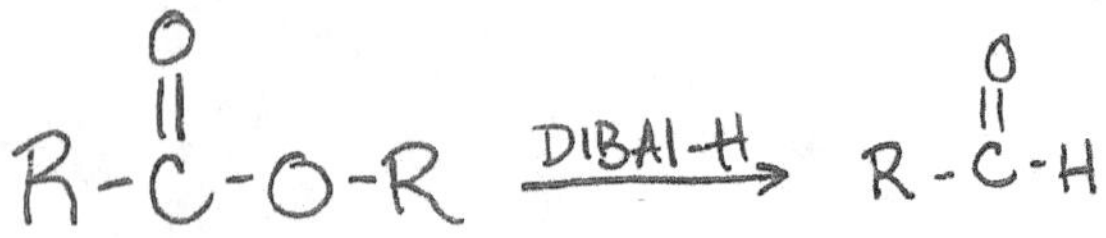

$$R-\overset{\overset{\displaystyle O}{||}}{C}-O-R \xrightarrow{\text{DIBAl-H}} R-\overset{\overset{\displaystyle O}{||}}{C}-H$$

$$R-\overset{\overset{\displaystyle O}{||}}{C}-Cl \xrightarrow{\text{DIBAl-H}} R-\overset{\overset{\displaystyle O}{||}}{C}-H$$

Specific Examples:

Key points:
- Reducing agent
- High reactivity
- DIBAL-H is called diisobutyl aluminum hydride (see structure below):

III. Alkanes

1. <u>Catalytic hydrogenation of an alkene or alkyne</u>

Functional Group Transformation: Alkene → Alkane

General Scheme:

$$R-CH = CH + H_2 \xrightarrow{Pt./c} R-CH_2CH_3$$

Specific Example:

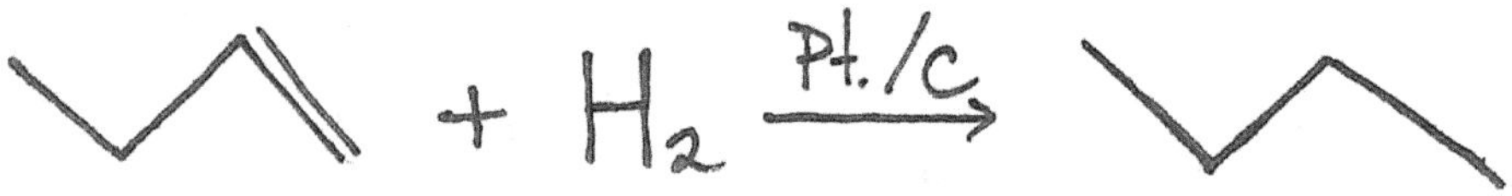

Key points:
- H_2 with Platinum (Pt/C) as a catalyst,
- H_2 with Palladium (Pd/C) as a catalyst or
- H_2 with Nickel (Ni) as a catalyst
- Syn addition
- <u>No reduction (NR) to esters, amides, or carboxylic acids</u> when using H_2.

2. <u>Wolff-Kishner & Clemmensen reduction of an aldehyde or a ketone</u>

Functional Group Transformation: carbonyl group → methylene group

General Scheme:

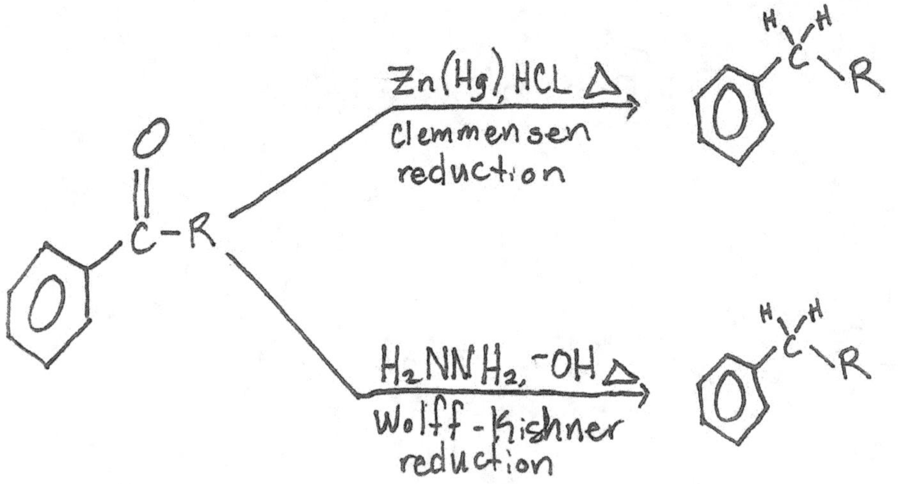

Specific Example:

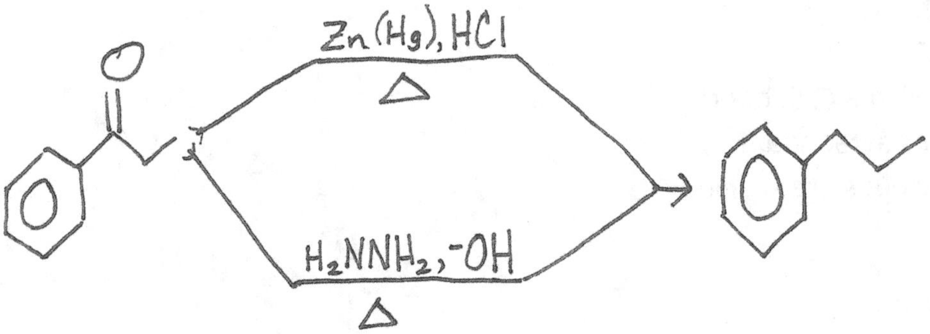

Key points:
- Heat required for reaction (Δ)
- Oxygen removed from the reactant (deoxygenation)
- Benzene ring unaffected

3. Reaction of a Gilman reagent with an aryl or alkyl halide

Functional Group Transformation: Alkyl halide → Alkane

General Scheme:

$$R_2\text{-}CuLi + R'\text{-}X \longrightarrow R\text{-}R' + R\text{-}Cu + LiX$$

Specific Example:

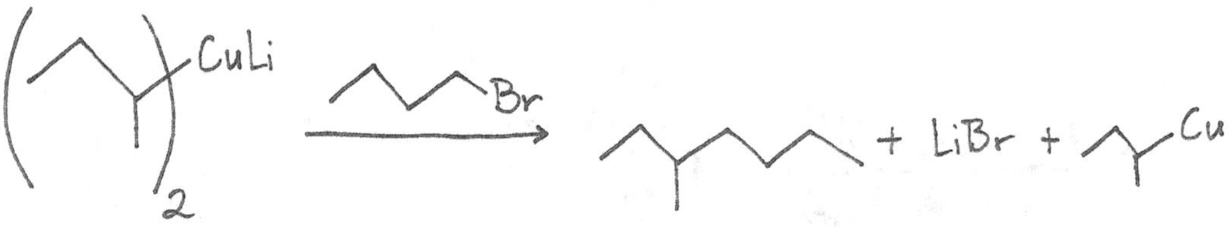

Key points:
- Forming a new C-C bond
- $2R\text{-}Li + CuI \rightarrow R\text{-}CuLi$
- Replaces C-X bond with a C-C bond
- Must have 2 R-groups attached
- See below how 2 R-groups are formed:

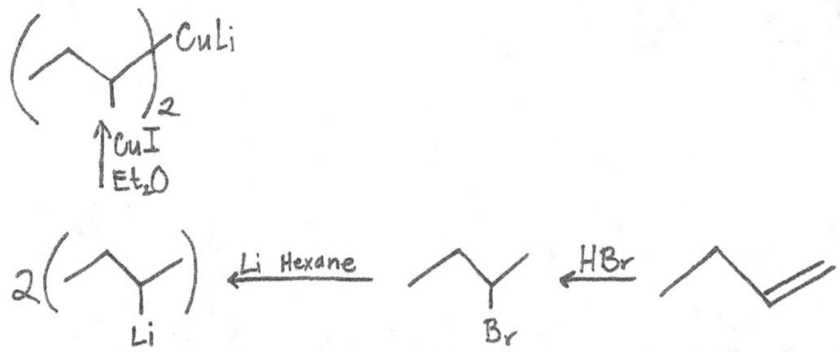

4. Reaction of a Grignard reagent with a source of protons

Functional Group Transformation: Alkyl halide → Alkane

General Scheme:

$$R-CH_2-CH-R' \xrightarrow{\text{Mg}}_{\text{THF}} R-CH_2-CH-R' \xrightarrow{\text{H}_2\text{O}}_{\text{H}\oplus} R-CH_2-CH_2-R'$$

$$\underset{X}{\big|} \qquad\qquad \underset{Mg-X}{\big|}$$

Specific Example:

Key points:
- X = Halide (i.e. Br, Cl)
- Removes halide and replaces it with a hydrogen
- **Cannot be prepared from compounds that contain acidic groups**
- Diethyl ether or THF (tetrahydrofuran) used as a solvent

IV. Alkenes

1. **Hydrogenation of an alkyne with Lindlar's catalyst → cis alkene**

Functional Group Transformation: alkyne → alkene

General Scheme:

$$R-C\equiv C-R + H_2 \xrightarrow{\text{Lindlar's Cat.}} \underset{R}{\overset{H}{>}}C=C\underset{R}{\overset{H}{<}}$$

Specific Example:

$$\text{———} + H_2 \xrightarrow{\text{Lindlar's Cat.}} \text{⌐—⌐}\quad cis$$

Key points:
- Syn addition
- Forms a cis alkene
- Poisoned catalyst stops after one H_2 addition

2. <u>Reduction of an alkyne with Na (or Li) and liq. NH3 → trans alkene</u>

Functional Group Transformation: alkyne → alkene

General Scheme:

$$R-C\equiv C-R + H_2 \xrightarrow[\substack{NH_3(\ell) \\ -78°C}]{Na}$$

Specific Example:

$$+ H_2 \xrightarrow[\substack{NH_3(\ell) \\ -78°C}]{Na}$$

Trans

Key points:
- Trans addition
- One electron transfer mechanism

3. Elimination of hydrogen halide from an alkyl halide

Functional Group Transformation: alkyl halide → alkene

General Scheme:

$$R - \underset{\underset{R}{|}}{\overset{\overset{R}{|}}{C}} - \ddot{\underset{..}{X}}: \ + \ ^{\ominus}\!\overset{..}{\underset{..}{O}}H \longrightarrow CH_2 = C \overset{R}{\underset{R}{\diagup}} + H_2\overset{..}{O}: + :\overset{..}{\underset{..}{X}}:^{\ominus}$$

Specific Example:

$$ \times^{\overset{..}{\underset{..}{Br}}:} + \ ^{\ominus}\!\overset{..}{\underset{..}{O}}H \longrightarrow \lambda + H_2\overset{..}{O}: + :\overset{..}{\underset{..}{Br}}:^{\ominus}$$

Mechanism:

$$CH_3 - \overset{\overset{CH_3}{|}}{\underset{\underset{:\overset{..}{Br}:}{|}}{\overset{\alpha}{C}}} - _{\beta}CH_2^{\overset{H}{}} + ^{\ominus}\!\overset{..}{\underset{..}{O}}H \longrightarrow CH_3 - \overset{\overset{CH_3}{|}}{C} = CH_2 + H_2\overset{..}{O}: + :\overset{..}{\underset{.}{Br}}:^{\ominus}$$

Key points:
- E2 reaction
- Alpha carbon where halogen is located
- Beta carbon where hydrogen is eliminated and double bond is formed (may have more possibilities depending on the substituent of that beta carbon)
- Also called 1,2-elimination reaction
- Need anti-coplanarity

4. Other elimination of hydrogen halide from an alkyl halide

Functional Group Transformation: alkyl halide → alkene

General Schemes:

$$\boxed{E2} \quad H_3C-CH_2-X + {}^{\ominus}\!:\!\ddot{O}H \xrightarrow{\;E2\;} H_2C=CH_2 + H_2\ddot{O}: + :\ddot{X}:^{\ominus}$$

$$\boxed{E1} \quad H_3C-\overset{\overset{\displaystyle R}{|}}{\underset{\underset{\displaystyle R}{|}}{C}}-X + H_2\ddot{O}: \xrightarrow{\;E1\;} H_2C=C\overset{R}{\underset{R}{\diagdown}} + H_3O^{\oplus} + :\ddot{X}:^{\ominus}$$

Specific Examples:

$$\boxed{E2} \quad \triangle_{Br} + {}^{\ominus}\!:\!\ddot{O}H \xrightarrow{\;E2\;} H_2C=CH_2 + H_2\ddot{O}: + :\ddot{B}r:^{\ominus}$$

$$\boxed{E1} \quad H_3C-\overset{\overset{\displaystyle R}{|}}{\underset{\underset{\displaystyle R}{|}}{C}}-X + H_2\ddot{O}: \xrightarrow{\;E1\;} H_2C=C\overset{R}{\underset{R}{\diagdown}} + H_3O^{\oplus} + :\ddot{X}:^{\ominus}$$

Mechanisms:

$$\ominus:\!\ddot{O}H \; + \; H-\overset{H}{\underset{H}{C}}-CH_2-\overset{..}{Br} \;\longrightarrow\; H_2C=CH_2 + H_2\ddot{O}: + :\!\ddot{Br}:^{\ominus}$$

$$E1 \qquad H_3C-\overset{CH_3}{\underset{CH_3}{C}}-\overset{..}{Br} \;\rightleftharpoons\; H_3C-\overset{CH-H}{\underset{CH_3}{C}}\!\!^{\oplus} \; + :\!\ddot{Br}:^{\ominus} \quad \xrightarrow{H_2\ddot{O}:} \quad \overset{H_3C}{\underset{H_3C}{C}}\!\!=CH_2 + H_3O^{\oplus}$$

Key points:

- No carbocation intermediate for E2
- E1 forms a carbocation intermediate
- Beta hydrogens adjacent to alkyl halide abstracted in both cases

5. Formation of a cyclic alkene using a Diels-Alder reaction

Functional Group Transformation: conjugated diene and dienophile → cyclic alkene

General Scheme:

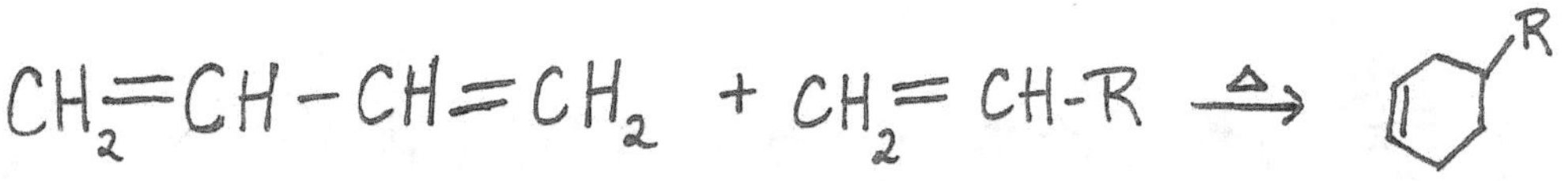

$$CH_2=CH-CH=CH_2 \ + \ CH_2=CH\text{-}R \ \xrightarrow{\triangle} \ $$

Specific Example:

Mechanism:

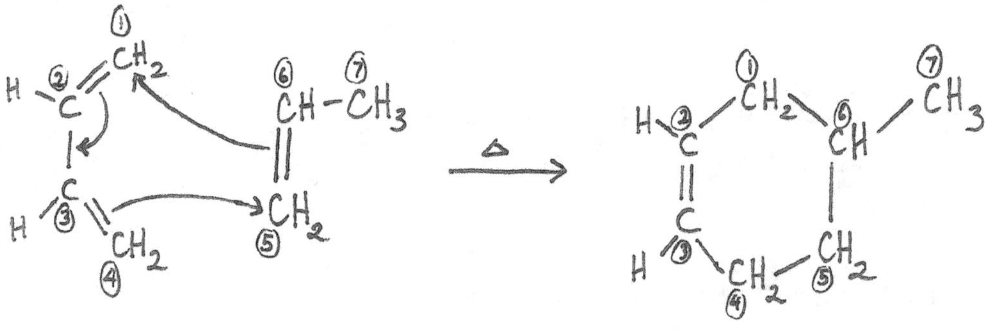

Key points:
- Concerted reaction
- Boat transition state
- Pericyclic reaction

6. Acid-catalyzed dehydration of an alcohol

Functional Group Transformation: Alcohol → Alkene

General Scheme:

$$R-CH_2-\underset{\underset{OH}{|}}{CH}-R \; \underset{\Delta}{\overset{H_2SO_4}{\rightleftharpoons}} \; R-CH=CH-R \; + \; H_2O \,(dilute)$$

Specific Example:

$$\underset{\Delta}{\overset{H_2SO_4}{\rightleftharpoons}} \qquad + \; H_2O \; (dilute)$$

Mechanism:

Key points:
- Works best for 2° (secondary) and 3° (tertiary) alcohols
- Carbocation intermediate
- E1
- Beta hydrogen abstracted
- Heat required (Δ)
- Important, the **major product is the more substituted alkene!**
- i.e. CH₃–CH=CH–CH₃ > CH₃–CH=CH₂ (the fewer "H" on alkene, the better)

7. Reaction of a Gilman reagent with a halogenated alkene

Functional Group Transformation: halogenated alkene → alkene

General Scheme:

Specific Examples:

Key points:
- Can be used when S_N2 cannot
- Forming new C-C bond
- Et_2O or THF (THF is used under anhydrous conditions)
- Must have 2 R-groups attached
- See bottom of page 18 on how 2 R-groups are formed

8. <u>Alkene formed from tert-butoxide (-O-tBu or T-Buo-)</u>

Functional Group Transformation: bromoethylbenzene → styrene

General Scheme:

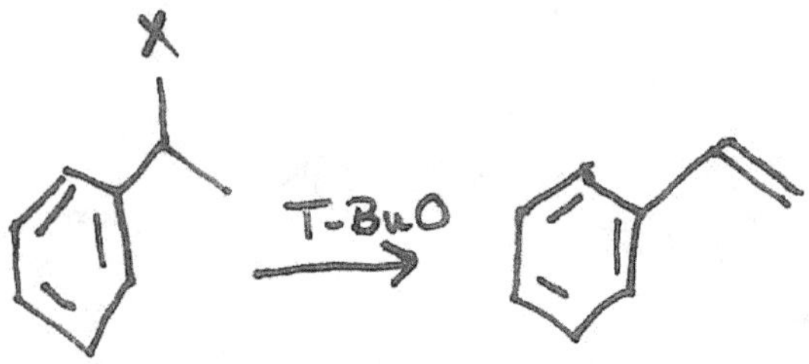

Specific Examples:

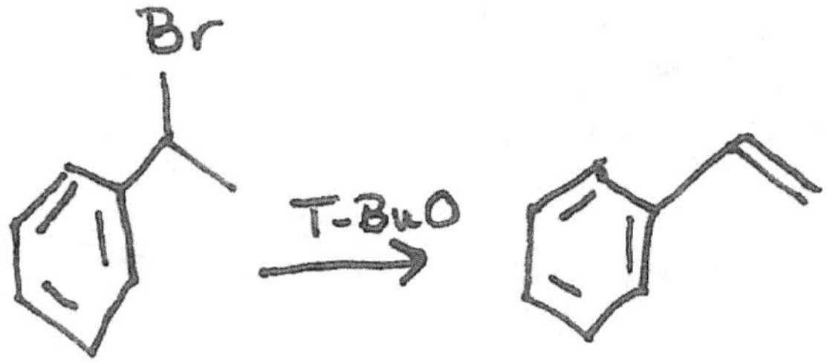

Key points:
- Removes the halide to form a double bond
- Ethylbenzene can form into bromoethylbenzene when NBS w/(Δ) is added (see page 64)

V. Alkyl Halides

1. Addition of hydrogen halide (HX) to an alkene

Functional Group Transformation: alkene → alkyl halide

General Scheme:

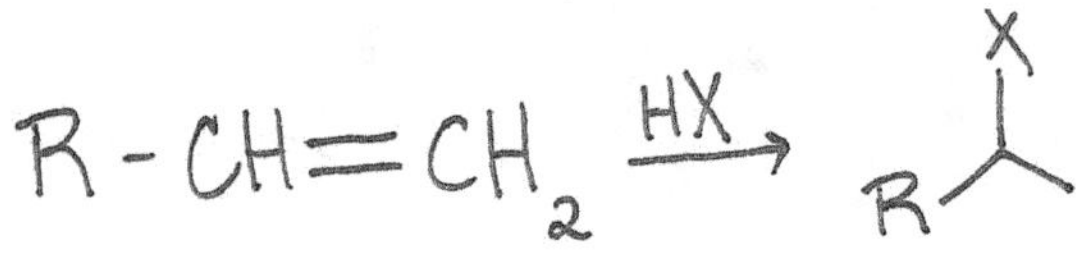

$$R - CH = CH_2 \xrightarrow{HX} R \overset{X}{\underset{}{\bigvee}}$$

Specific Example:

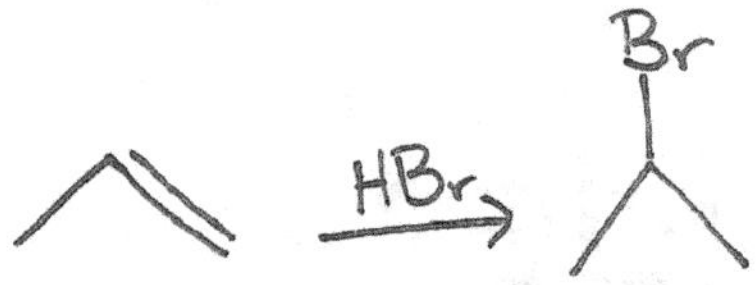

Mechanism:

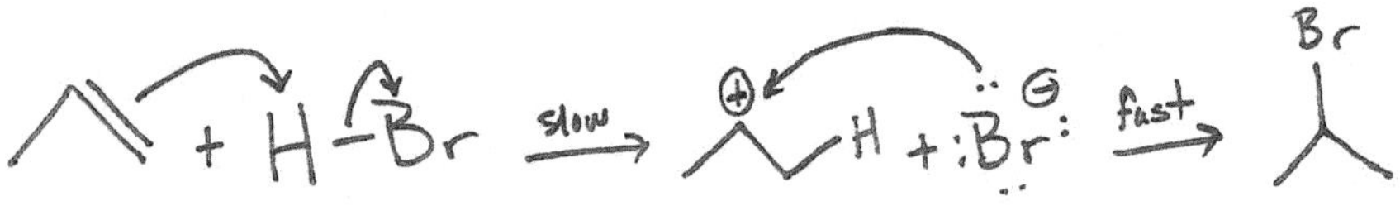

Key points:
- X = Br, Cl, I
- Markovnikov Addition
- Cation rearrangement (hydride or alkyl shift) if a more stable carbocation results
- Halide anion can approach equally from both faces of sp_2 carbocation → racemic mixture

2. Addition of halogen (X₂) to an alkene

Functional Group Transformation: alkene → alkyl halide

General Scheme:

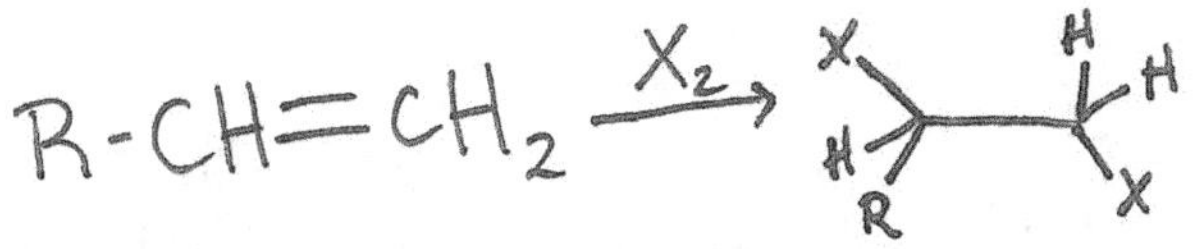

Specific Examples:

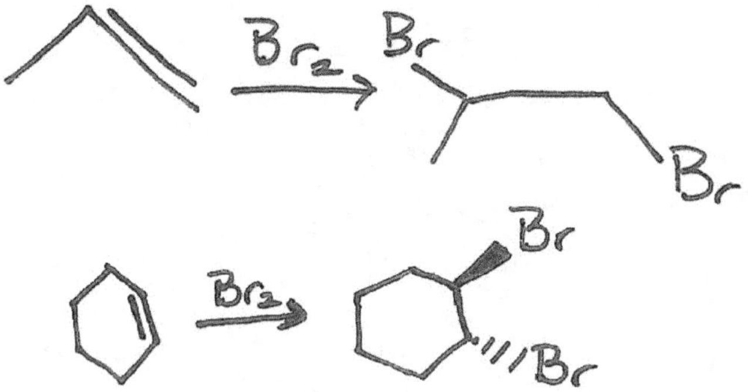

Mechanism:

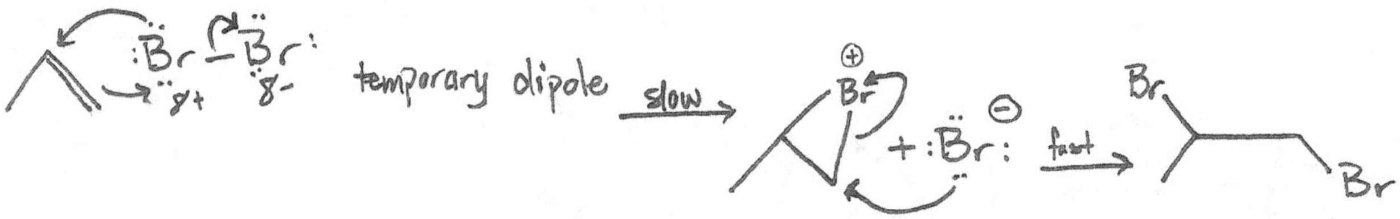

Key points:
- $X = Br, Cl$
- No rearrangement since no carbocation intermediate
- Anti-Markovikov addition (AMA)
- Catalyst not required
- Can use an inert solvent such as dichloromethane (CH_2Cl_2) but it does not participate in the reaction

3. Addition of hydrogen halide or halogen to an alkyne

Functional Group Transformation: alkyne → alkyl halide

General Scheme:

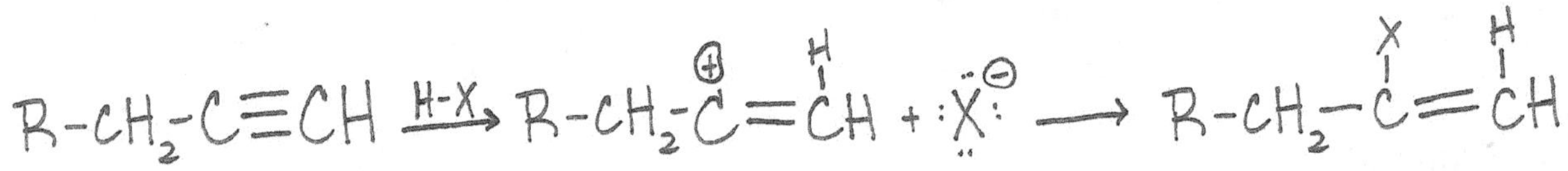

$$R-CH_2-C\equiv CH \xrightarrow{H-X} R-CH_2-\overset{\oplus}{C}=\overset{H}{C}H + :\overset{..}{\underset{..}{X}}:^{\ominus} \longrightarrow R-CH_2-\overset{X}{C}=\overset{H}{C}H$$

Specific Example:

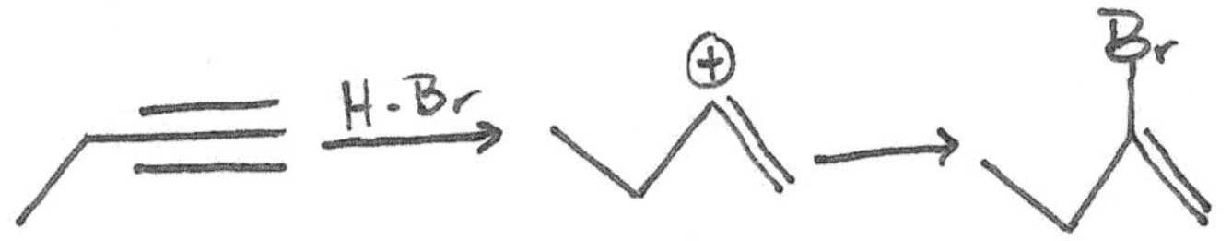

Mechanism:

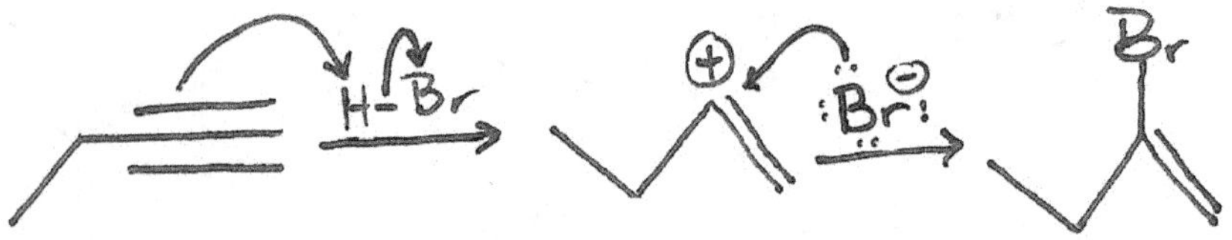

Key points:
- Markovnikov addition
- Carbocation intermediate

4. Reaction of an alcohol with hydrogen halide, SOCl₂, PCl₃, or PBr₃

Functional Group Transformation: alcohol → alkyl halide

General Schemes:

$$R\text{-}OH + PBr_3 \xrightarrow{Pyr.} R\text{-}Br$$

$$R\text{-}OH + PCl_3 \xrightarrow{Pyr.} R\text{-}Cl$$

$$R\text{-}OH + SOCl_2 \xrightarrow{Pyr.} R\text{-}Cl$$

Specific Examples:

(drawings of alcohols reacting with PBr₃, PCl₃, SOCl₂ to form alkyl halides, and a secondary alcohol reacting with PBr₃ and SOCl₂)

Key points:
- Pyridine (pyr.) used. See below:

- Back side attack (get inversion at carbon) for PBr₃
- For SOCl₂, you can get retention or inversion (mostly retention)
- Pyridine generally used as the solvent to prevent the buildup of HBr or HCl

5. <u>Cleavage of an ether with HI or HBr</u>

Functional Group Transformation: ether → alkyl halide

General Scheme:

$$R-O-R' + H-X \xrightarrow[H_2O]{\Delta} R-OH + R'-X \rightarrow R-X + R'-X + H_2O$$
(excess)

Specific Example:

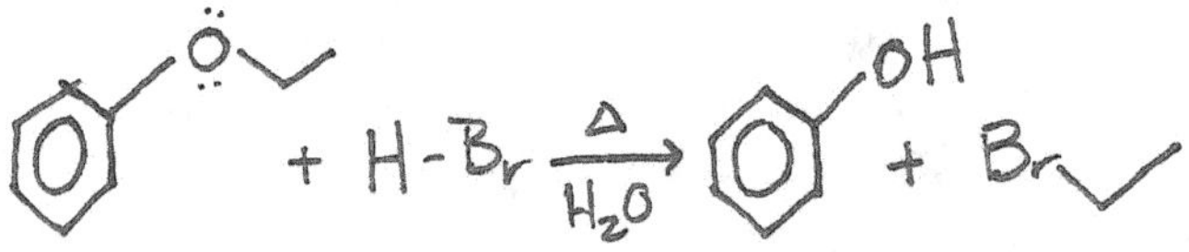

Key points:
- Cleaved by strong acids
- Heat required (Δ)

6. <u>Alkane with halogens in the presence of light (hv)</u>

Functional Group Transformation: alkane → alkyl halide

General Scheme:

$$R-CH_3 + X_2 \xrightarrow{hv} R-CH_2X$$

Specific Example:

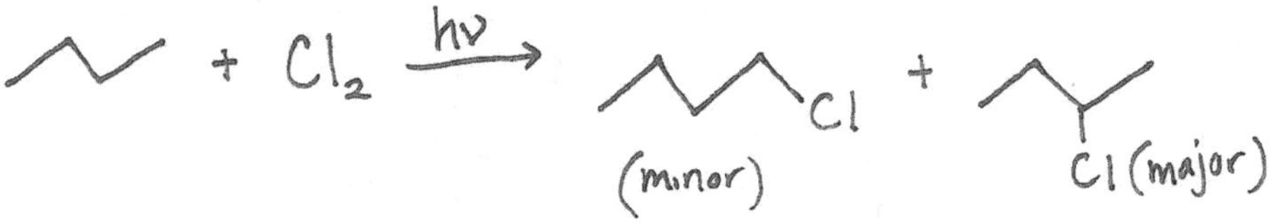

Key points:
- Rate of formation increases: <u>tertiary (3^0) > secondary (2^0) > primary (1^0)</u>
- hv = Irradiation with light
- For mechanism see page 64

VI. Alkynes

1. **Reaction of an acetylide ion with an alkyl halide**

Functional Group Transformation: acetylide ion → alkyne

General Scheme:

$$R-CH_2-C\equiv C{:}^{\ominus}\,Na^{\oplus} + R-CH_2-X \longrightarrow R-CH_2-C\equiv C-CH_2-R + {:}\ddot{X}{:}^{\ominus}$$

Specific Example:

Mechanism:

$$CH_3-CH_2-C\equiv C{:}^{\ominus} + CH_3CH_2CH_2{-}Br \longrightarrow CH_3CH_2C\equiv C-CH_2CH_2CH_3 + {:}\ddot{B}r{:}^{\ominus}$$

$$\uparrow Na^{\oplus}NH_2^{\ominus}$$

$$CH_3-CH_2-C\equiv C-H$$

Key points:
- Nucleophile (negatively charged acetylide ion) is attracted to the partially positively charged carbon (electrophile) of the alkyl halide
- <u>Only</u> **primary alkyl halides or methyl halides** should be used in this reaction
- $NaNH_2$ removes proton on alkyne

2. <u>Elimination of hydrogen halide from an alkene</u>

Functional Group Transformation: alkene → alkyne

General Scheme:

$$R\text{-}CH=CH\text{-}R \xrightarrow[CH_2Cl_2]{X_2} R\text{-}\overset{X}{\underset{X}{CH}}\text{-}\overset{}{CH}\text{-}R \xrightarrow{NaNH_2} R\text{-}CH=\overset{}{\underset{X}{CH}}\text{-}R \xrightarrow[\;]{\overset{+\,NaX}{NaNH_2}} R\text{-}C\equiv C\text{-}R + NaX$$

Specific Example:

Key points:
- E2 elimination reaction
- X = halide (i.e. Br, I)
- <u>Strong base</u> (i.e. NaNH$_2$) is needed for the second elimination
- If using a <u>weaker base</u> (-OH) at room temperature, reaction will stop at vinylic halide only
- Dichloromethane used (CH$_2$Cl$_2$)

3. <u>Two successive eliminations of hydrogen halide from a geminal dihalide</u>

Functional Group Transformation: alkyl dihalide → alkyne

General Scheme:

$$R-CH_2-\overset{\overset{\displaystyle X}{|}}{\underset{\underset{\displaystyle X}{|}}{C}}-R \xrightarrow{2NaNH_2} RC\equiv CR \ + \ 2NH_3 + 2NaX$$

Specific Example:

$$\xrightarrow{2NaNH_2} \underline{\equiv} \ + \ 2NH_3 \ + \ 2NaBr$$

Key points:
- Two halogens are on the same carbon (geminal dihalides)
- X = halide (i.e. Br, I)
- Must have a very strong base (i.e. NaNH$_2$)

VII. Amines

1. Reduction of a nitro compound

Functional Group Transformation: nitro substituent → amino substituent

General Scheme:

$$\text{Ph—NO}_2 \xrightarrow[\text{2. NaOH}]{\text{1. Sn, HCl}} \text{Ph—NH}_2$$

Specific Examples:

$$\text{Ph—NO}_2 \xrightarrow[\text{2. NaOH}]{\text{1. Sn, HCl}} \text{Ph—NH}_2 + H_2O$$

$$\text{Ph—NO}_2 \xrightarrow{\text{Pt, H}_2} \text{Ph—NH}_2$$

Key points:
- Option a) 1.Sn, HCl /2. NaOH
- Option b) Pt./H_2 (Pt. = platinum)

2. Reaction of an alkyl halide with -NH₂

Functional Group Transformation: alkyl halide → substitution product (amine)

General Scheme:

Specific Example:

Mechanism:

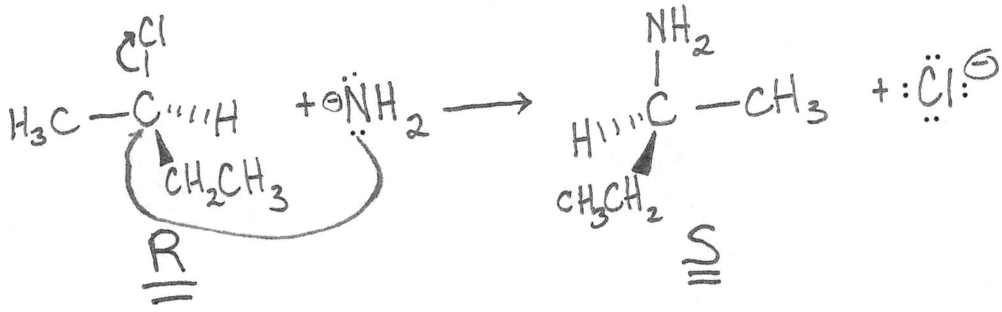

Key points:

- Concerted mechanism
- Primary or secondary alkyl halides for $S_{N}2$
- Back side attack
- Configuration change in most cases (check priority) from "R" to "S" or from "S" to "R"

3. Reaction of a Gabriel synthesis

Functional Group Transformation: alkyl halide → primary amine

General Scheme:

Specific Example:

Mechanism:

Key points:
- Converts a primary alkyl halide into a primary amine

VIII. Carboxylic Acids

1. Oxidation of an alkyl benzene

Functional Group Transformation: alkyl Benzene → carboxylic acid

General Scheme:

$$\text{Ar-R} \xrightarrow[\text{2. H}^{\oplus}]{\text{1. KMnO}_4, \Delta} \text{Ar-COOH}$$

$$\xrightarrow[\text{2. H}^{\oplus}]{\text{1. KMnO}_4, \Delta} \text{No Reaction}$$
(No benzylic Hydrogen)

Specific Example:

$$\text{Ph-CH}_3 \xrightarrow[\text{2. H}^{\oplus}]{\text{1. KMnO}_4, \Delta} \text{Ph-CO}_2\text{H}$$

$$\xrightarrow[\text{2. H}^{\oplus}]{\text{1. KMnO}_4, \Delta} \text{N.R.}$$

Key points:
- 1. KMnO$_4$, 2. H+ and heat (commonly used)
- Must have a hydrogen on benzylic

2. Oxidation of a primary alcohol

Functional Group Transformation: alcohol → carboxylic acid

General Scheme:

$$R-CH_2-OH \xrightarrow{[O]} R-COOH$$

Specific Example:

Key points:
- [O] = $KMnO_4$ or Jones oxidation

IX. 1,2-Diol

1. Reaction of an epoxide with water

Functional Group Transformation: epoxide → 1,2-diol

General Scheme:

$$\text{epoxide} \xrightarrow{H^{\oplus}, H_2O} HO-CH_2CH_2-OH$$

Specific Example:

$$\xrightarrow{H^{\oplus}, H_2O} HO\diagdown\diagup OH$$

Mechanism:

$$HO-CH_2-CH_2 \rightleftharpoons HO-CH_2-CH_2-OH + H_3O^{\oplus}$$

$$+ H_2O:$$

Key points:

- Can be opened via acid catalysis or base catalysis:

Acid catalysis: $\xrightarrow[H_2O]{H^+}$

Base catalysis: $\xrightarrow{NaOH}$

X. Epoxides

1. Reaction of an alkene with a peroxyacid

Functional Group Transformation: alkene → epoxide

General Scheme:

$$RCH=CH_2 + \underset{(Peroxyacid)}{RC\overset{O}{\overset{\|}{}}OOH} \longrightarrow RCH\overset{\overset{O}{\diagup\diagdown}}{-}CH_2 + \underset{(Carboxylic\ acid)}{RC\overset{O}{\overset{\|}{}}OH}$$

Specific Example:

$$\wedge + \xrightarrow{mCPBA} \triangle\!\!\!-O + CH_3C\overset{O}{\overset{\|}{}}OH$$

Key points:
- Concerted mechanism
- mCPBA (meta-chloroperoxybenzoic acid) most common peroxyacid
- Notice the slight difference of peroxyacid vs. carboxylic acid
- mCPBA:

XI. Esters

1. Recation of a sulfonate ester with alcohol

Functional Group Transformation: alcohol → sulfonate ester

General Sceme:

$$R-OH + Cl-\overset{O}{\underset{O}{\overset{||}{\underset{||}{S}}}}-R' \xrightarrow{\text{Pyr}} RO-\overset{O}{\underset{O}{\overset{||}{\underset{||}{S}}}}-R' + :\overset{..}{\underset{..}{Cl}}:^{\ominus} + \text{Pyr}{-}H^{\oplus}$$

Specific Example:

$$CH_3OH + Cl-\overset{O}{\underset{O}{\overset{||}{\underset{||}{S}}}}-CH_3 \xrightarrow{\text{Pyr}} CH_3{-}O-\overset{O}{\underset{O}{\overset{||}{\underset{||}{S}}}}-CH_3 + :\overset{..}{\underset{..}{Cl}}:^{\ominus} + \text{Pyridinium}^{\oplus}{-}H$$

Key points:

- Pyridine (pyr.) prevents build up of HCl
- Examples of R' (mesylate, tosylate, and triflate)

  $R' = CH_3$ mesylate

  $R' = $ Tosylate

  $R' = CF_3$ Triflate

XII. Ethers

1. Acid-catalyzed addition of an alcohol to an alkene

Functional Group Transformation: alkene → ether

General Scheme:

$$RCH=CH_2 + CH_3OH \xrightarrow{H_2SO_4} RCH-CH_3$$
$$\underset{\displaystyle OCH_3}{|}$$

Specific Example:

$$\diagup\!\!\!\!\diagup + CH_3OH \xrightleftharpoons{H_2SO_4} \diagup\!\!\!\!\diagdown O-CH_3$$

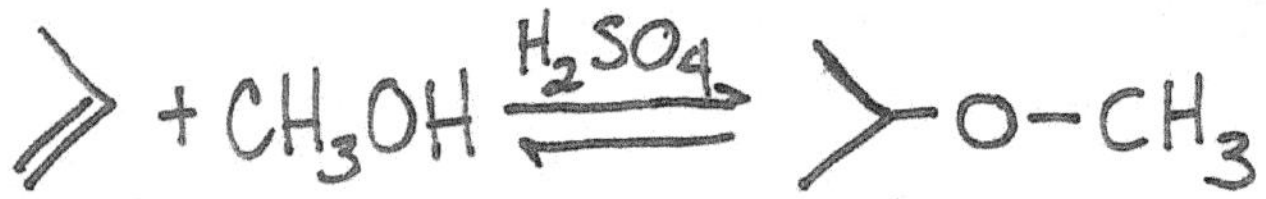

Mechanism:

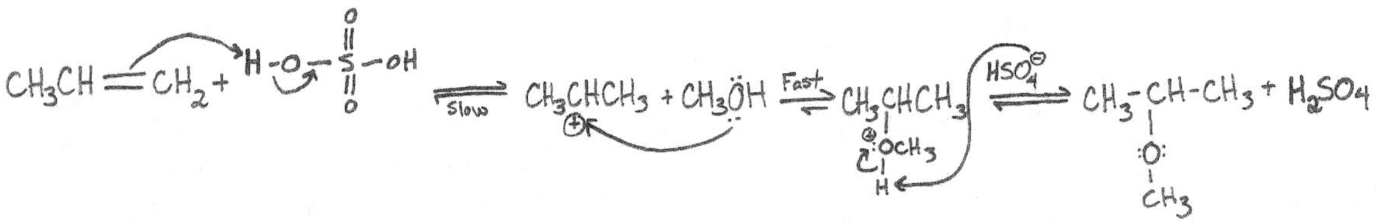

Key points:
- Requires an acid catalyst (H_2SO_4)
- Markovnikov addition

2. <u>Alkoxymercuration-demercuration of an alkene</u>

Functional Group Transformation: alkene → ether

General Scheme:

Specific Example:

Key points:
- Markovnikov addition

3. <u>Williamson ether synthesis: reaction of an alkoxide ion with an alkyl halide</u>

Functional Group Transformation: alkyl halide → ether

General Scheme:

$$R-X + R-\ddot{\underset{\cdot\cdot}{O}}{:}^{\ominus} \longrightarrow ROR + :\ddot{\underset{\cdot\cdot}{X}}{:}^{\ominus}$$

$$\uparrow_{NaH}$$

$$R-\ddot{O}H$$

Specific Example:

$$CH_3-Br + CH_3\ddot{\underset{\cdot\cdot}{O}}{:}^{\ominus} \longrightarrow CH_3-O-CH_3 + :\ddot{\underset{\cdot\cdot}{Br}}{:}^{\ominus}$$

$$\uparrow_{NaH}$$

$$CH_3\ddot{\underset{\cdot\cdot}{O}}H$$

Key points:
- $S_{N}2$ reaction
- NaH needed to remove a proton from an alcohol (deprotonation)

4. Formation of symmetrical ethers by heating an acidic solution of a primary alcohol

Functional Group Transformation: primary alcohol → ether

General Scheme:

$$R-OH + R'-\overset{\oplus}{O}H_2 \xrightarrow{SN2} R-\overset{\oplus}{\underset{|}{O}}-R' \longrightarrow R-O-R' + H\overset{\oplus}{B}$$

with $\uparrow H^+$ and $R'-OH$ below, and $B:\nearrow$ attacking H

Specific Example:

Mechanism:

Key points:
- Protonation of the most basic atom
- Back side attack by the nu: (nucleophile)
- Substitution product
- S_N2
- B: = Base (i.e. H_2O)

XIII. Halogenation

1. Bromination & Chlorination in the presence of light (hv) or NBS

Functional Group Transformations: Ethylbenzene (w/NBS) → alkyl halide

cyclohexane → alkyl halide

General Schemes:

Specific Examples:

Mechanism:

Key points:
- hv = Irradiation with light forming free radicals
- Increasing rate (preference) as follows: 3° radical > 2° radical > 1° radical
- Caution: NBS works when benzylic hydrogen is present (see page 48 for benzylic example)
- Heat required when NBS is used
- Three steps involved: initiation step, propagation steps, and termination steps

XIV. Halohydrins

1. <u>Reaction of an alkene with Br_2 (or Cl_2) and H_2O</u>

Functional Group Transformation: alkene → halohydrins

General Scheme:

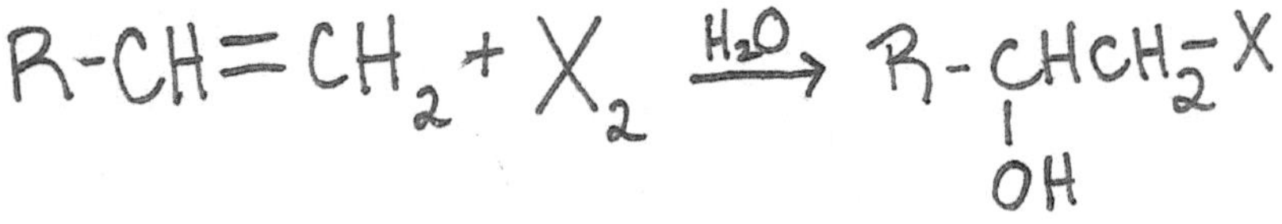

$$R-CH=CH_2 + X_2 \xrightarrow{H_2O} R-\underset{\underset{OH}{|}}{C}HCH_2-X$$

Specific Example:

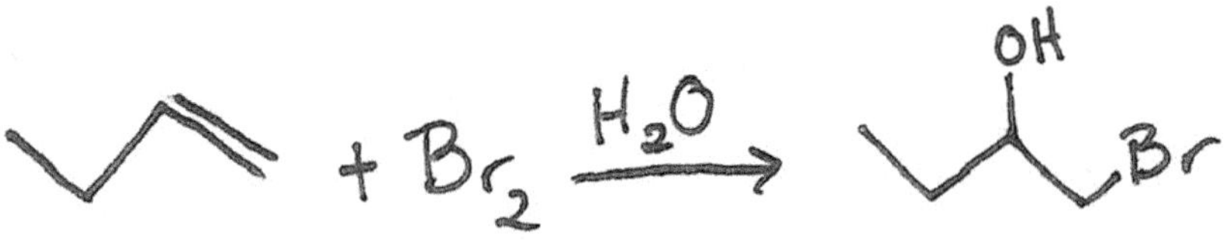

Mechanism:

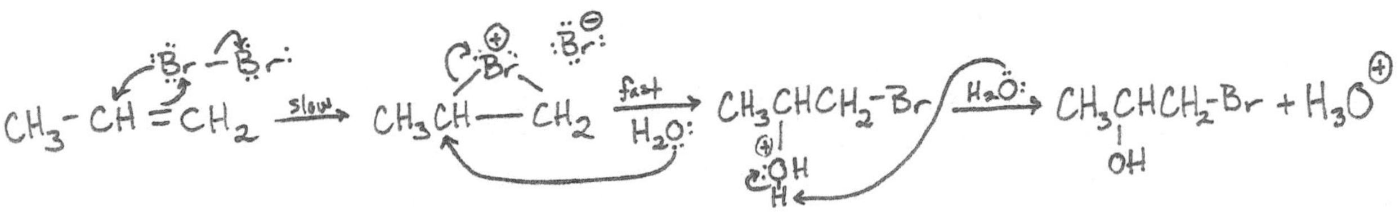

Key points:
- Major product includes OH
- Halide (Br) attaches to carbon with more hydrogens (Markovnikov addition) during first step (not -OH)
- If CH_2Cl_2 is used as a solvent (<u>not H_2O</u>), Bromine will be in place of OH! (See page 33)

2. Reaction of an epoxide with a hydrogen halide

Functional Group Transformation: epoxide → halohydrins

General Scheme:

$$H_2C \underset{\displaystyle \overset{\ddot{O}}{\triangle}}{} CH_2 + H\text{-}X \longrightarrow HO\text{-}CH_2\text{-}CH_2\text{-}X$$

Specific Example:

$$\triangle\ddot{O} + H\text{-}Br \longrightarrow HO\diagdown\!\diagup Br$$

Mechanism:

$$H_2C \overset{\ddot{O}}{\triangle} CH_2 + H\text{-}\ddot{Br} \longrightarrow H_2C \overset{\overset{\oplus}{O}H}{\triangle} CH_2 \quad :\ddot{Br}:^{\ominus} \longrightarrow HO\diagdown\!\diagup Br$$

Key points:
- X = Br, I, Cl
- Epoxide react with nu: (nucleophile) under acidic conditions to give ring-opened alcohols
- Acid catalyst assists epoxide ring opening by providing a better LG (leaving group) alcohol at the site of nu: attack

XV. Ketones

1. <u>Addition of water to an alkyne</u>

Functional Group Transformation: terminal alkyne → ketone

General Schemes:

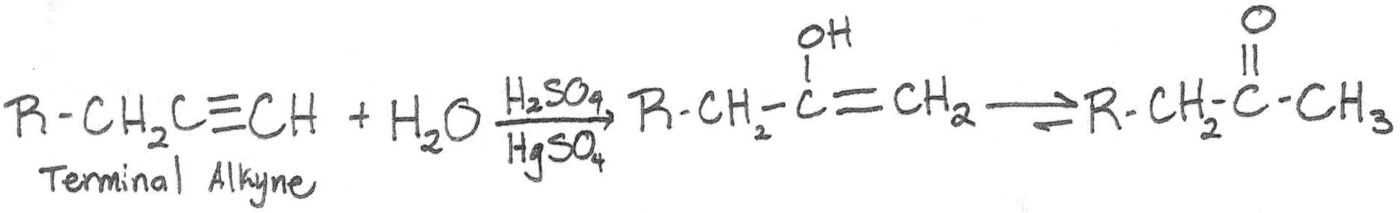

$$R-CH_2C\equiv CH + H_2O \xrightarrow[HgSO_4]{H_2SO_4} R-CH_2-\overset{\overset{OH}{|}}{C}=CH_2 \rightleftharpoons R-CH_2-\overset{\overset{O}{||}}{C}-CH_3$$

Terminal Alkyne

Specific Examples:

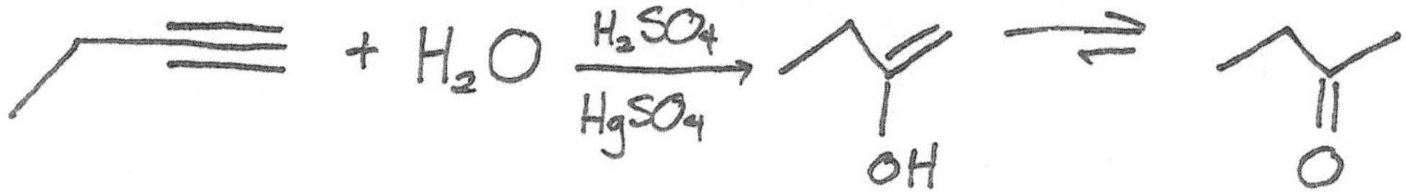

Key points:
- Catalyst needed (H_2SO_4)
- During 2nd intermediate, Markovnikov addition is used
- Enol formed
- End product formed after tautomerization

2. <u>Hydroboration-oxidation of an alkyne</u>

Functional Group Transformation: internal alkyne → ketone

General Scheme:

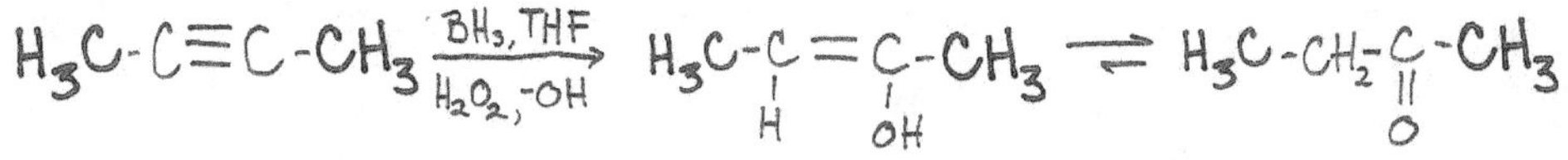

$$H_3C-C\equiv C-CH_3 \xrightarrow[H_2O_2,-OH]{BH_3,THF} H_3C-\underset{H}{\underset{|}{C}}=\underset{OH}{\underset{|}{C}}-CH_3 \rightleftharpoons H_3C-CH_2-\underset{O}{\overset{||}{C}}-CH_3$$

Specific Example:

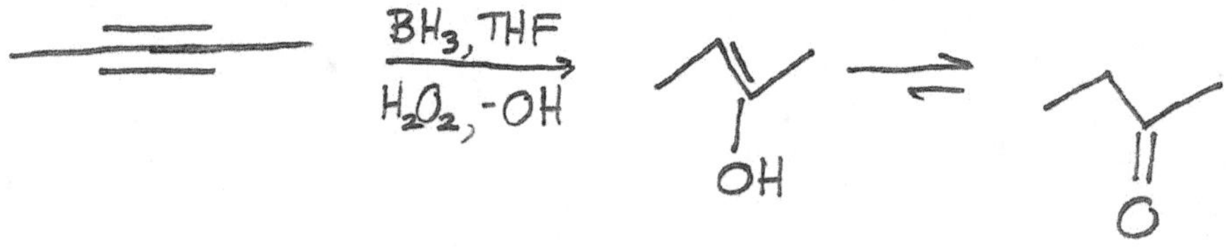

Key points:
- Anti-Markovnikov addition
- Enol formed

3. Friedel-Crafts acylation

Functional Group Transformation: benezene ring → ketone

General Scheme:

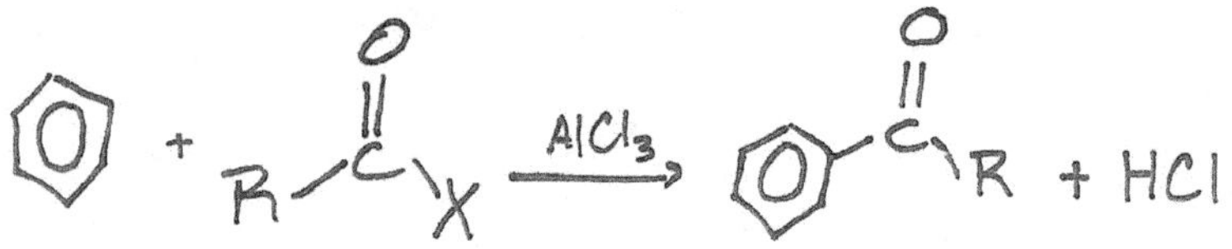

Specific Example:

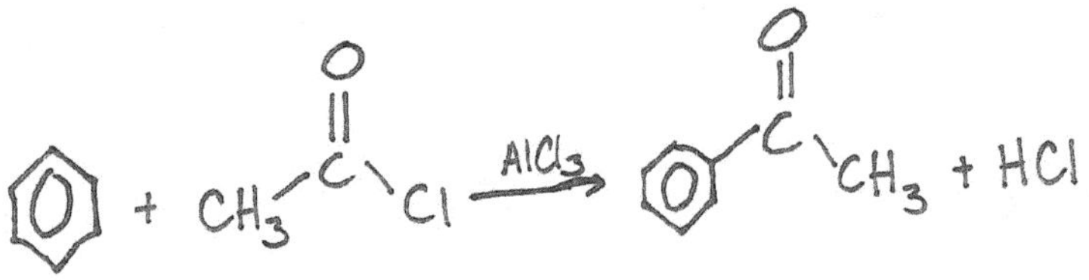

Mechanism:

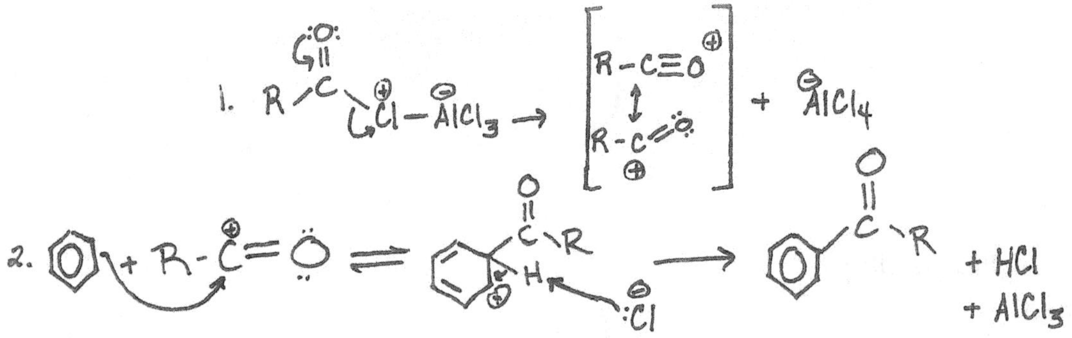

Key points:
- New C-C bond
- Base = -Cl

4. Oxidation of a secondary alcohol

Functional Group Transformation: secondary alcohol → ketone

General Schemes:

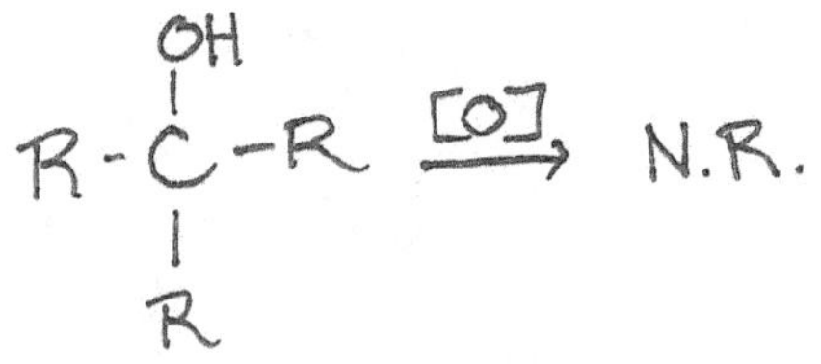

Specific Examples:

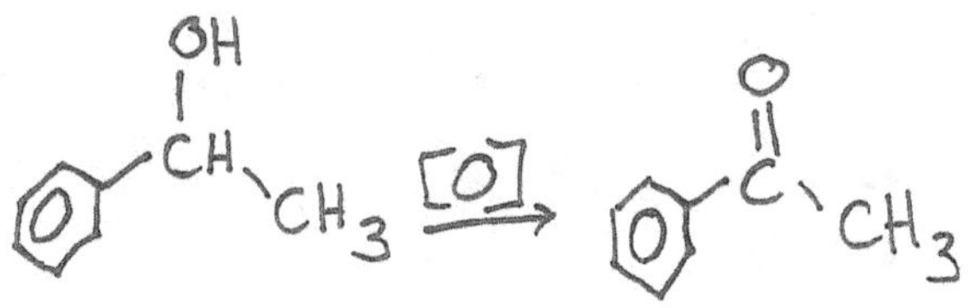

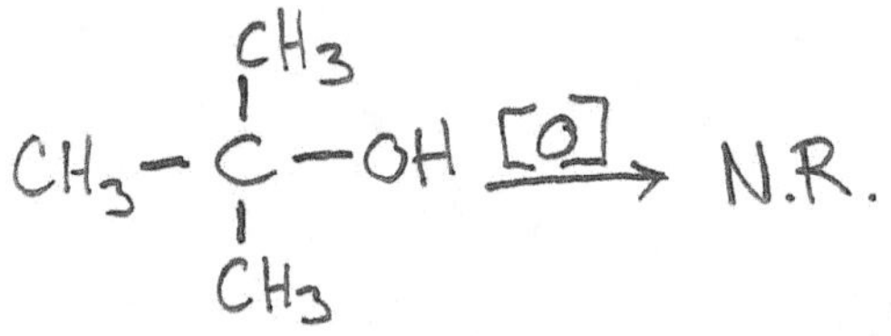

Key points:
- An <u>alpha carbon must have a hydrogen</u> in order for a reaction to occur
- [O] = Cr(VI), PCC (w/CH_2Cl_2), or $KMnO_4$ (very strong)

XVI. Nitrile

1. Reaction of an alkyl halide with cyanide ion

Functional Group Transformation: alkyl halide → nitrile

General Scheme:

$$R-CH_2-\underset{\underset{R}{|}}{CH}-X + NaC\equiv N \longrightarrow R-CH_2-\underset{\underset{R}{|}}{CH}-C\equiv N + :\overset{..}{\underset{..}{I}}:^{\ominus} \quad Na^{\oplus}$$

Specific Example:

Mechanism:

Key points:
- S_N2
- Concerted mechanism
- No carbocation intermediate
- Primary or secondary for S_N2
- Back side attack
- Inverted configuration in most cases (check priority) from "S" to "R" or from "R" to "S"

XVII. Substituted Benezenes

1. <u>Halogenation</u>

Functional Group Transformation: benezene ring → halogenation

General Scheme:

Specific Example:

Mechanism:

Key points:
- X = halide (Br, Cl)
- Fe used as Lewis Acid catalyst

2. Nitration

Functional Group Transformation: benezene ring → nitration

General Scheme:

Specific Example:

Nitric acid   Nitrobenzene

Mechanism:

1.

2.

Key points:
- Acid catalyzed reaction

3. <u>Sulfonation</u>

Functional Group Transformation: benezene ring → sulfonation

General Scheme:

benzene $+ H_2SO_4 \overset{\Delta}{\rightleftharpoons}$ C_6H_5-SO_3H $+ H_2O$

Specific Example:

benzene $+ H_2SO_4 \overset{\Delta}{\rightleftharpoons}$ C_6H_5-SO_3H $+ H_2O$

Mechanism:

1. $H\ddot{O}-SO_3H + H-OSO_3H \rightleftharpoons H\overset{\oplus}{\ddot{O}}-SO_3H + HSO_4^{\ominus} \rightleftharpoons \overset{\oplus}{SO_3H}$
     with H below

2. $+ \overset{\oplus}{SO_3H} \overset{\Delta}{\rightleftharpoons}$ $C_6H_5(SO_3H)(H)^{\oplus} \xrightarrow{HSO_4^{\ominus}}$ $C_6H_5-SO_3H + H_2SO_4$

Key points:
- Reaction is reversible
- Heat required to reverse reaction

4. Friedel-Crafts acylation

Functional Group Transformation: benezene ring → Friedel-Crafts acylation

General Scheme:

Specific Example:

Mechanism:

Key points:
- New C-C bond

5. Friedel-Crafts alkylation

Functional Group Transformation: benezene ring → Friedel-Crafts alkylation

General Scheme:

$$\bigcirc + RCl \xrightarrow{AlCl_3} \bigcirc^R + HCl$$

Specific Example:

$$\bigcirc + CH_3Cl \xrightarrow{AlCl_3} \bigcirc^{CH_3} + HCl$$

Mechanism:

1. $R-\ddot{C}l: + AlCl_3 \longrightarrow R-\overset{\oplus}{C}l-\overset{\ominus}{A}lCl_3$

2. $\bigcirc \ R-\overset{\oplus}{C}l-\overset{\ominus}{A}lCl_3 \longrightarrow \bigcirc^R H \xrightarrow{Cl-AlCl_3} \bigcirc^R + HCl + AlCl_3$

Key points:
- New C-C bond
- AlBr$_3$ also works (in sub. of AlCl$_3$) when halide on R group is Br

6. <u>Sandmeyer reaction</u> : reaction of an arenediazonium salt w/CuBr, CuBI, or CuCN

Functional Group Transformation: arenediazonium salt → sandmeyer reaction

General Scheme:

Specific Examples:

Benzenediazonium bromide   bromobenzene p-toluenediazonium chloride   p-chlorotoluene

m-bromobenzenediazonium chloride   m-bromobenzonitrile

Key points:
- Copper is needed
- Formation of nitrogen gas will result
- Cl− or Br− ion present

7. <u>Formation of a phenol by reaction of an Arenediazonium salt and acid</u>

Functional Group Transformation: arenediazonium salt → phenol

General Scheme:

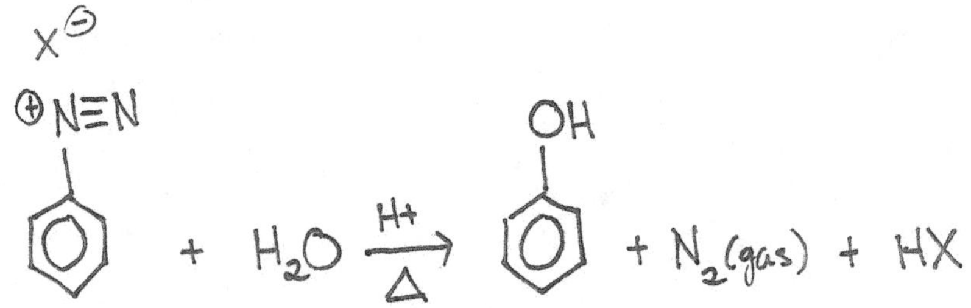

Specific Example:

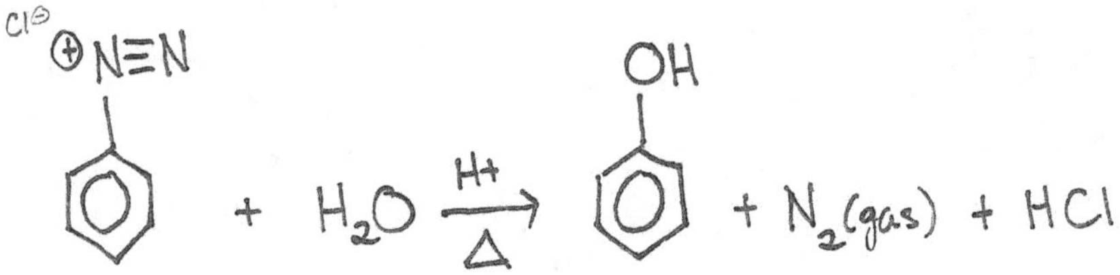

Key points:
- Heat required
- Acid catalyst needed
- Nitrogen gas released

8. **Formation of an aniline by reaction of a benzyne intermediate with $-NH_2$**

Functional Group Transformation: benzyne → aniline

General Scheme:

Specific Example:

Mechanism:

1.

2.

or

Key points:
- -NH$_2$ removes a β-proton
- Benzyne intermediate formed
- Protonation of the resulting anion forms the substitution product
- 50/50 mixture of constitutional isomers

XVIII. Sulfide

1. Reaction of a sulfide

Functional Group Transformation: dimethyl sulfide → sulfonium salt

General Scheme:

$$R - \ddot{S} - R' + R'' - \ddot{X}: \xrightarrow[SN2]{Pyr.} R - \overset{\oplus}{\underset{\underset{R''}{|}}{\ddot{S}}} - R + \ominus\ddot{X}:$$

Specific Example:

$$\diagup \ddot{S} \diagdown + CH_3 - \ddot{\underset{..}{I}}: \xrightarrow[SN2]{Pyr.} \diagup \overset{\oplus}{\underset{|}{\ddot{S}}} \diagdown + :\overset{\ominus}{\underset{..}{I}}:$$

Mechanism:

$$H_3C - \ddot{S} - CH_3 + H_3C - \overset{\curvearrowright}{\ddot{I}}: \xrightarrow[SN2]{Pyr.} CH_3 - \overset{\oplus}{\underset{\underset{CH_3}{|}}{\ddot{S}}} - CH_3 + :\overset{\ominus}{\underset{..}{I}}:$$

Key points:
- S_N2 reaction
- Configuration change in most cases
- Check priorities if chiral
- Has a weakly basic leaving group
- Pyridine (pyr.) used as the solvent (see below):

XIX. Thiol

1. Reaction of a thiol with an alkyl halide

Functional Group Transformation: alkyl halide → thiol

General Scheme:

$$R-\ddot{X}: \; + \; H\ddot{S}:^{\ominus} \longrightarrow R-\ddot{S}H \; + :\ddot{X}:^{\ominus}$$

Specific Example:

$$\text{—}\ddot{B}r: \; + :\overset{\ominus}{\ddot{S}}\text{-}H \longrightarrow \text{—}\ddot{S}H \; + :\ddot{B}r:^{\ominus}$$

Mechanism:

$$CH_3CH_2-\ddot{B}r: \; + \; H\ddot{S}:^{\ominus} \xrightarrow{\;SN2\;} CH_3CH_2-\ddot{S}H \; + :\ddot{B}r:^{\ominus}$$

ethanethiol

Key points:
- S_N2
- Back side attack
- Configuration change in most cases (check priorities if chiral)
- Ethyl bromide becomes ethanethiol. See mechanism

XX. Summaries and Examples

Elimination and Substitution Summaries

Substrates	Base/Nucleophile	Products	Mechanism
Primary (1°) R – X	Strong **Hindered** Base (i.e. tert-butoxide)	E2 Major, S_N2 Minor	E2, S_N2
Secondary (2°) X | R – CH – R′	Strong **Hindered** Base (i.e. tert-butoxide)	E2 High Majority, S_N2 Trace	E2 Highly Preferred
Tertiary (3°) X | R – C – R′ | R″	Strong **Hindered** Base (i.e. tert-butoxide)	E2 Major, E1 Minor, S_N1 Trace	E2 Highly Preferred

Substrates	Base/Nucleophile	Products	Mechanism
Primary (1°) R – X	Strong **Unhindered** Base (i.e. -NH2, -OR, -OH)	S_N2 Major, E2 Minor	S_N2, E2
Secondary (2°) X | R – CH– R′	Strong **Unhindered** Base (i.e. -NH2, -OR, -OH)	S_N2 & E2 (_more E2_ in **protic solvent!**)	S_N2, E2
Tertiary (3°) X | R – C – R′ | R″	Strong **Unhindered** Base (i.e. -NH2, -OR, -OH)	E2 Major, E1 Minor, S_N1 Minor	E2 Highly Preferred

Substrates	Base/Nucleophile	Products	Mechanism
Primary (1°) R – X	Weak Base/Good Nucleophile (i.e. -Br, -Cl, CN-, RS-)	S_N2 Majority, E2 Trace	S_N2 Highly Preferred
Secondary (2°) X | R – CH– R′	Weak Base/Good Nucleophile (i.e. -Br, -Cl, CN-, RS-)	S_N2 Major, E2 Minor	S_N2, E2

Substrates	Base/Nucleophile	Products	Mechanism
Secondary (2°) X | R – CH– R′	Weak Base (Solvent such as H_2O, CH_3OH etc.)	S_N1 Major, E1 Minor	S_N1 Highly Preferred, E1
Tertiary (3°) X | R – C – R′ | R″	Weak Base (Solvent such as H_2O, CH_3OH etc.)	S_N1 Major, E1 Minor	S_N1 Highly Preferred, E1

Substrates	Base/Nucleophile	Products	Mechanism
Methyl – X CH3 – X	Any Base	Only S_N2	Only S_N2

Direct addition of H_2

Reactant (H2)	Usual Catalyst(s)	Reduction Product
Acyl Chloride O || R-C-Cl	Pd/C	R – CH2 – OH Primary alcohol
Acyl Chloride	Lindlar's catalyst	Aldehyde
Nitrile (R–C≡N)	Pt or Pd/C	Primary amine R-CH2-NH2
Imine R-CH=NR'(H)	Pt or Pd/C	R-CH2-NH2(R')
Alkene or alkyne	Pt, Pd, Ni	Alkane
Alkyne	Lindlar's catalyst	Alkene (cis)
Aldehyde	Raney Ni Pd/C	Primary alcohol
Ketone	Raney Ni Pd/C	Secondary alcohol

Different multi-step reaction examples
(putting it all together!)

The next several examples are just one way (out of many) to get to the end product. Test yourself by looking at the starting reagent(s) and end product and see if you can figure out the steps in between then see what was written. Please note each step on every answer key has page numbers for reference. Think of them like a puzzle. Enjoy.

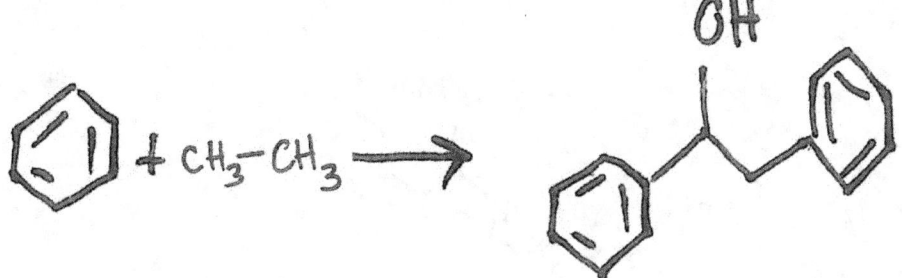

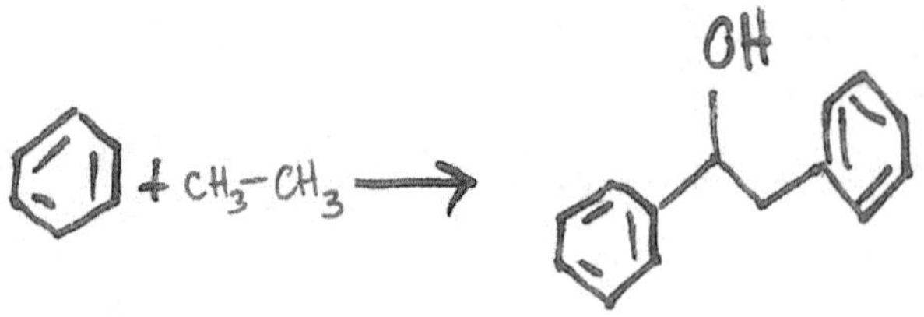

OH

⬡ + CH₃–CH₃ ⟶ (1-phenyl-2-phenylethanol structure with OH)

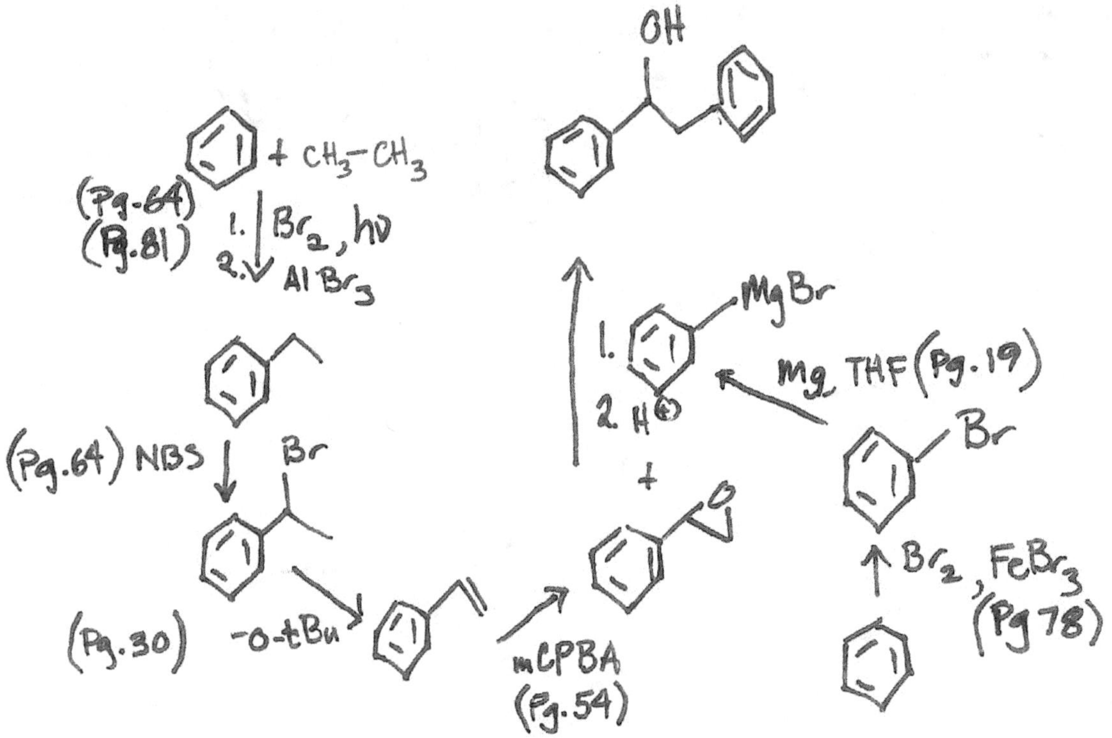

OH

⬡ + CH₃–CH₃

(Pg.64)
(Pg.81)   1. | Br₂, hν
          2. ↓ AlBr₃

(Pg.64) NBS ↓   Br

1.
2. H⊕

MgBr

Mg THF (Pg.19)

Br

(Pg.30) –O-tBu

+

mCPBA
(Pg.54)

↑ Br₂, FeBr₃
   (Pg.78)

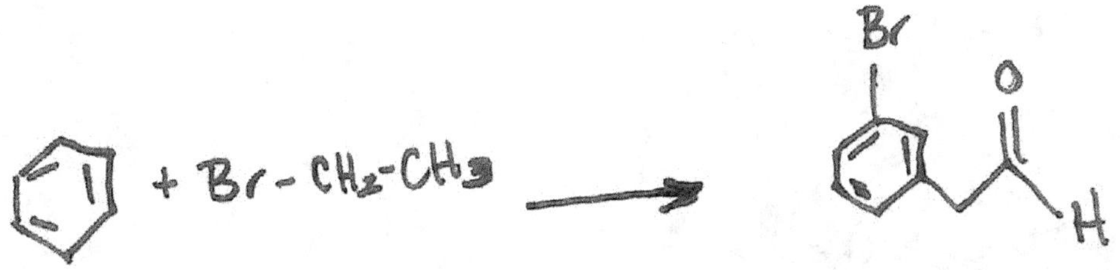
+ Br-CH2-CH3
Br
O
H

benzene + Br-CH₂-CH₃ ⟶ (3-bromophenyl)acetaldehyde

benzene + Br-CH₂-CH₃

(Pg. 81) ↓ AlBr₃

ethylbenzene

(Pg. 64) 1. | NBS, ROOR Δ
(Pg. 30) 2. | T-BuO⁻

styrene

(Pg. 33) 1. | Br₂, CH₂Cl₂
(Pg. 42) 2. | 2NaNH₂

phenylacetylene

H₂SO₄
Hg SO₄   ↘
(Pg. 70)   acetophenone

Br₂
FeBr₃
(Pg. 78) → 3-bromoacetophenone

H₂NNH₂,
-OH, Δ (Pg. 17) ↗

1-bromo-3-ethylbenzene

1. NBS, ROOR Δ (Pg. 64)
2. T-BuO⁻   (Pg. 30) ↑

3-bromostyrene

1. Br₂, CH₂Cl₂ (Pg. 33)
2. 2NaNH₂   (Pg. 42) ↑

3-bromophenylacetylene

1. sia₂BH
2. H₂O₂, -OH (Pg. 12) ↑

(3-bromophenyl)acetaldehyde

$$\text{\raisebox{-2pt}{⌒}} + CH_3-CH_3 + CH_4 \longrightarrow$$

$CH_3-CH_3 + CH_4 \longrightarrow$

(target structure: methyl cyclohexene)

CH_3-CH_3

(Pg.64) $\downarrow$ Br_2, $h\nu$

$\downarrow$ Br_2, $h\nu$,

CH_3-CH_2-Br

Br

(Pg.30) $\downarrow$ T-BuO$^{\ominus}$

$\downarrow$ T-BuO$^{\ominus}$

$\uparrow (CH_3)_2 CuLi$ $\xleftarrow{CuI}$ $CH_3^{\ominus} Li$

(Pg.18)

Br

$\uparrow$ Li, Hexane   Pg.18

CH_3-Br

$CH_2=CH_2$

$+$ $\xrightarrow[\text{(Pg.27)}]{\text{Diels Alder}}$

(Pg.64) $\downarrow$ NBS $\triangle$

$\nearrow$ NBS $\triangle$

$\uparrow$ Br_2, $h\nu$ (Pg.64)

CH_4

Br

$+$ T-BuO$^{\ominus}$ $\rightarrow$

$$CH_3-CH_3 \ + \ \text{[benzene]} \ \longrightarrow \ \text{[3-phenylpropanal]}$$

CH_3-CH_3 + [benzene] $\longrightarrow$ [3-phenylpropanal structure with $\overset{O}{\underset{H}{}}$]

[benzene] $\xrightarrow[\substack{AlBr_3 \\ (Pg.82)}]{CH_3CH_2Br}$ [ethylbenzene] $\xrightarrow[\substack{\Delta}]{\substack{(Pg.64) \\ NBS}}$ [$\overset{Br}{|}$ benzylic bromide] $\xrightarrow[(Pg.30)]{t.BuO^{\ominus}}$ [styrene]

H-Br (up arrow to first benzene)

$CH_2=CH_2$
1. $Br_2, h\nu$ (Pg.64)
2. T-BuO$^{\ominus}$ (Pg.30)

CH_3-CH_3

[styrene] $\xrightarrow[\substack{2. H_2O_2, HO^{\ominus}}]{\substack{1. BH_3, THF \ (Pg.4)}}$ [phenethyl alcohol —OH]

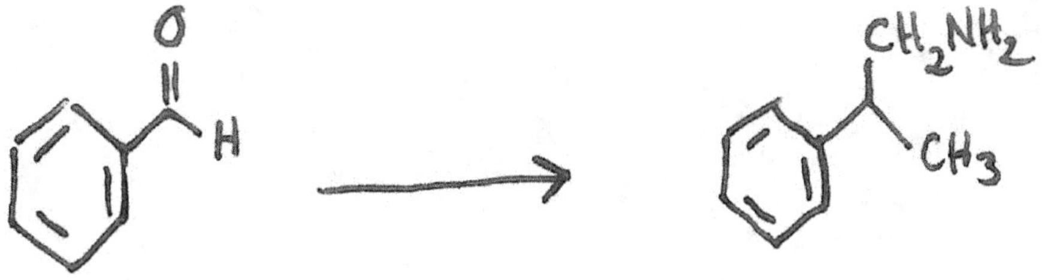
O
H
CH2NH2
CH3

Starting material: benzaldehyde (PhCHO) → product: CH_2NH_2 / CH_3 substituted benzene (amphetamine-like structure)

Retrosynthesis / forward synthesis scheme:

Benzaldehyde (PhCHO)

(Pg.17) ↓ Zn(Hg), HCl Δ
        or
        $H_2N\ddot{N}H_2$, $\bar{O}H$, Δ

→ toluene (Ph–CH_3)

(Pg.64) ↓ NBS, Δ / Peroxide

→ Ph–CH_2Br (benzyl bromide)

CuI + (CH$_3$)$_2$CuLi
2CH$_3$Li   (Pg.18)
↑ 4Li, hexanes
2CH$_3$Br ← 2CH$_4$
(Pg.64) Br_2, hv

→

NBS, Δ (pg.64)
Peroxide ↑

→ ethylbenzene (Ph–CH$_2$CH$_3$)

→ 1-bromoethylbenzene (Ph–CHBr–CH$_3$)

(Pg.76) NaCN ↑

→ Ph–CH(C≡N)–CH$_3$

(Pg.16) 2H$_2$, Pt. ↑

→ product: CH_2NH_2 / CH_3 substituted benzene

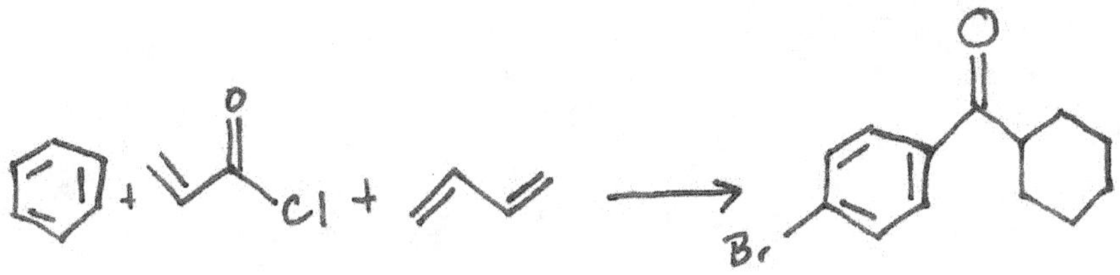
O
Cl
O
Br

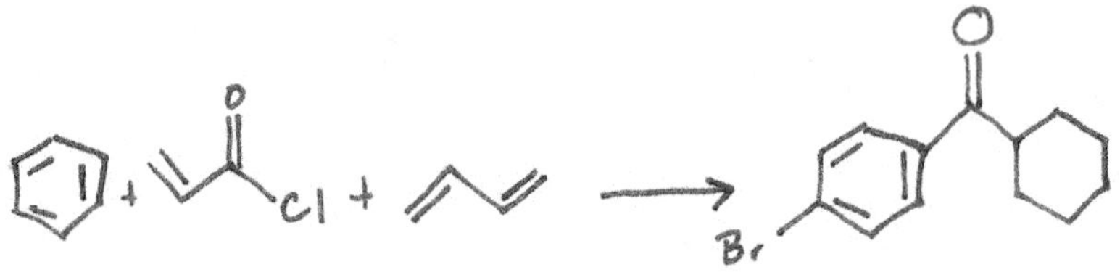

+ + ⟶

Br

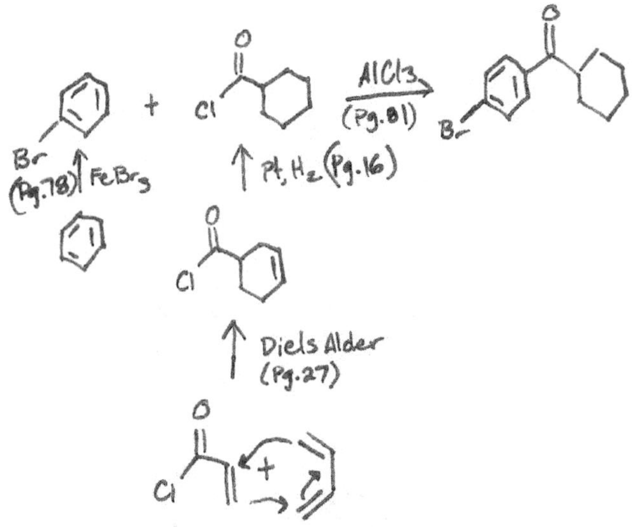

+ Cl $\xrightarrow{\text{AlCl}_3}$ Br
(Pg.81)

Br
(Pg.78) ↑ FeBr₃

↑ Pt, H₂ (Pg.16)

Cl

↑ Diels Alder
(Pg.27)

O
Cl

$$CH_4 \quad + \quad \text{[benzene]} \quad \longrightarrow \quad \text{[benzophenone]}$$

CH_4 + (benzene) → (benzophenone)

CH_4
$\downarrow$ Br_2 (Pg.64)
    $h\nu$

$\dfrac{1.CH_3Br}{2.AlBr_3}$ (Pg.82)

+

(benzene)

$\downarrow$

(toluene)

$\downarrow$ NBS, Pg.64
    ROOR, Δ

(benzyl bromide) CH_2-Br $\dfrac{1.\ 2Li, Hexane}{2.\ CuI, Et_2O}$ (Pg.18) → $\left(\text{(phenyl)} \right)_2 CuLi$ + (bromobenzene) Br

(diphenylmethanol) OH $\xrightarrow{\ PCC, CH_2Cl_2\ }$ (Pg.73)   (benzophenone) O

$-OH$ $\Big\uparrow$ NaOH
  or   DMSO (Pg.5)

Br
(bromodiphenylmethane)

$\Big\uparrow$ NBS, Δ (Pg.64)

(diphenylmethane)

$\Big\uparrow$

(Pg.78) $\Big\uparrow$ $\dfrac{Br_2}{FeBr_3}$

(benzene)

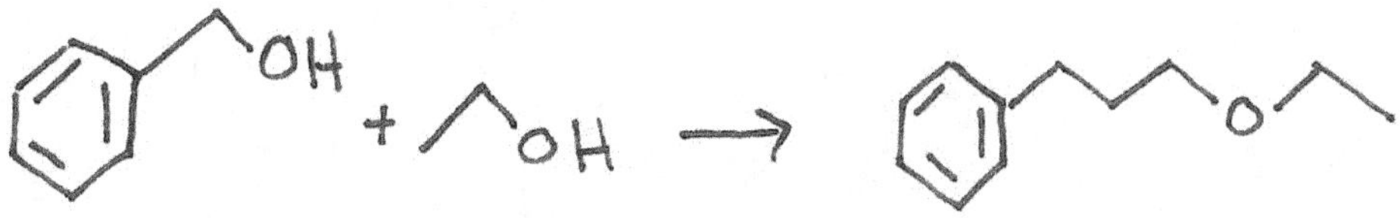

C₆H₅CH₂OH + CH₃CH₂OH → (phenyl-CH₂CH₂CH₂-O-CH₂CH₃)

Top reaction:

benzyl alcohol (◯–CH₂OH) + ∕∖OH → ◯–CH₂CH₂CH₂–O–CH₂CH₃

Synthesis scheme:

◯–CH₂OH
+
PBr₃
↓ Pyr. (Pg. 35)

◯–CH₂Br
↓ 2 Li (Pg. 18)
  Hexane

◯–CH₂Li + △O →[H⊕ (Pg. 52)] ◯–CH₂CH₂CH₂OH

CH₃–CH₂–OH
↓ H₂SO₄, Δ (Pg. 28)
CH₂=CH₂
↓ mCPBA (Pg. 54)
△O

Option 1 (Pg. 35) | PBr₃ Pyr.
↓
◯–CH₂CH₂CH₂Br
+
CH₃CH₂–Ö:⊖ ←[NaH (Pg. 60)] CH₃–CH₂–OH
↓
◯–CH₂CH₂CH₂–O–CH₂CH₃

Option 2 →[NaH (Pg. 60)] ◯–CH₂CH₂CH₂–Ö:⊖
+
∕∖Br ⇌[PBr₃ (Pg. 35) Pyr.] ∕∖OH
↓
◯–CH₂CH₂CH₂–O–CH₂CH₃

R—Br + NaCN   →   R—C≡N

$$R-COOH + SOCl_2 \rightarrow R-\overset{\overset{\displaystyle O}{\|}}{C}-Cl$$

R—C≡N + H₂O/H+   →   R—COOH

$$R-\overset{\overset{\displaystyle O}{\|}}{C}-Cl + H_2O \rightarrow R-COOH$$

R—C≡N + LiAlH₄   →   R—CH₂NH₂

$$R-\overset{\overset{\displaystyle O}{\|}}{C}-Cl + NH_3 \rightarrow R-\overset{\overset{\displaystyle O}{\|}}{C}-NH_2$$

$$R-\overset{\overset{\displaystyle O}{\|}}{C}-Cl + LiAlH_4 \rightarrow R-CH_2OH$$

$$R-C{\equiv}N + DIBAlH \rightarrow R-\overset{\overset{\displaystyle O}{\|}}{C}-H$$

$$R-\overset{\overset{\displaystyle O}{\|}}{C}-Cl + DIBAlH \rightarrow R-\overset{\overset{\displaystyle O}{\|}}{C}-H$$

$$R-C{\equiv}N + R'MgX \rightarrow R-\overset{\overset{\displaystyle O}{\|}}{C}-R'$$

$$R-\overset{\overset{\displaystyle O}{\|}}{C}-Cl + 2R'MgX \rightarrow R-\overset{\overset{\displaystyle R'}{|}}{\underset{\underset{\displaystyle R'}{|}}{C}}-OH$$

$$R-C{\equiv}N + SOCl_2 \rightarrow R-\overset{\overset{\displaystyle O}{\|}}{C}-Cl$$

$$R-\overset{\overset{\displaystyle O}{\|}}{C}-Cl + R'2CuLi \rightarrow R-\overset{\overset{\displaystyle O}{\|}}{C}-R'$$

$$R-\overset{\overset{\displaystyle O}{\|}}{C}-Cl + R'-COOH \rightarrow R-\overset{\overset{\displaystyle O}{\|}}{C}-O-\overset{\overset{\displaystyle O}{\|}}{C}-R'$$

Common reactions of Nitriles and Acyl Chlorides

$$R-\overset{\overset{\displaystyle O}{\|}}{C}-Cl + R'-OH \rightarrow R-\overset{\overset{\displaystyle O}{\|}}{C}-O-R'$$

$$R-\overset{\overset{\displaystyle O}{\|}}{C}-O-R' + H_2O/H+ \rightarrow R-COOH + R'-OH$$

$$R-\overset{\overset{\displaystyle O}{\|}}{C}-O-\overset{\overset{\displaystyle O}{\|}}{C}-R' + H_2O \rightarrow R-COOH$$

$$R-\overset{\overset{\displaystyle O}{\|}}{C}-O-R' + LiAlH_4 \rightarrow R-CH_2OH + R'-OH$$

$$R-\overset{\overset{\displaystyle O}{\|}}{C}-O-\overset{\overset{\displaystyle O}{\|}}{C}-R' + R'-OH \rightarrow R-\overset{\overset{\displaystyle O}{\|}}{C}-O-R'$$

$$R-\overset{\overset{\displaystyle O}{\|}}{C}-O-R' + DIBAlH \rightarrow R-\overset{\overset{\displaystyle O}{\|}}{C}-H$$

$$R-\overset{\overset{\displaystyle O}{\|}}{C}-O-R' + 2R'MgX \rightarrow R-\overset{\overset{\displaystyle R'}{|}}{\underset{\underset{\displaystyle R''}{|}}{C}}-OH$$

$$R-\overset{\overset{\displaystyle O}{\|}}{C}-O-\overset{\overset{\displaystyle O}{\|}}{C}-R' + NH_3 \rightarrow R-\overset{\overset{\displaystyle O}{\|}}{C}-NH_2$$

$$R-\overset{\overset{\displaystyle O}{\|}}{C}-O-R' + NH_3 \rightarrow R-\overset{\overset{\displaystyle O}{\|}}{C}-NH_2$$

===

$$R-\overset{\overset{\displaystyle O}{\|}}{C}-NH_2 + LiAlH_4 \rightarrow R-CH_2-NH_2$$

$$R-\overset{\overset{\displaystyle O}{\|}}{C}-NH_2 + H_2O/H+ \rightarrow R-COOH$$

$$R-\overset{\overset{\displaystyle O}{\|}}{C}-O-\overset{\overset{\displaystyle O}{\|}}{C}-R' + LiAlH_4 \rightarrow R-CH_2OH$$

===

Common reactions of Anhydrides, Esters, and Amides

$$R-\overset{\overset{\displaystyle O}{\|}}{C}-NH_2 + P_2O_5 \rightarrow R-C\equiv N$$

$$CH_3-\underset{\underset{CH_3}{|}}{\overset{\overset{CH_3}{|}}{C}}-CH=CH_2 \xrightarrow{Br_2, CH_2Cl_2} CH_3-\underset{\underset{CH_3}{|}}{\overset{\overset{CH_3}{|}}{C}}-\underset{\underset{Br}{|}}{CH}-CH_2-Br$$

$$CH_3-\underset{\underset{CH_3}{|}}{\overset{\overset{CH_3}{|}}{C}}-CH=CH_2 \xrightarrow{H_2O, H+} CH_3-\underset{\underset{OH}{|}}{\overset{\overset{CH_3}{|}}{C}}-\underset{\underset{CH_3}{|}}{CH}-CH_3$$

1,2 methyl shift

$$CH_3-\underset{\underset{CH_3}{|}}{\overset{\overset{CH_3}{|}}{C}}-CH=CH_2 \xrightarrow{HBr} CH3-\underset{\underset{CH_3}{|}}{\overset{\overset{Br}{|}}{C}}-\overset{\overset{CH_3}{|}}{CH}-CH_2$$

1,2 methyl shift

$$CH_3-\underset{\underset{CH_3}{|}}{\overset{\overset{CH_3}{|}}{C}}-CH=CH_2 \xrightarrow[\text{2.HO-, H}_2\text{O}_2, \text{H}_2\text{O}]{\text{1.BH}_3, \text{THF}} CH_3-\underset{\underset{CH_3}{|}}{\overset{\overset{CH_3}{|}}{C}}-CH_2-CH_2OH$$

Anti-Markovnikov (AMA)

$$CH_3-\underset{\underset{CH_3}{|}}{\overset{\overset{CH_3}{|}}{C}}-CH=CH_2 \xrightarrow{mCPBA} CH_3-\underset{\underset{CH_3}{|}}{\overset{\overset{CH_3}{|}}{C}}-\overset{O}{\overset{/\backslash}{CH-CH_2}}$$

$$CH3-\underset{\underset{CH_3}{|}}{\overset{\overset{CH_3}{|}}{C}}-CH=CH2 \xrightarrow[\text{2.HaBH}_4]{\text{1.Hg(OAc)}_2, \text{H}_2\text{O}, \text{THF}} CH_3-\underset{\underset{CH_3}{|}}{\overset{\overset{CH_3}{|}}{C}}-\underset{\underset{OH}{|}}{CH}-CH_3$$

No carbocation, no shift

Starting Alkenes using common reagents

Notes:

Notes:

Notes:

Notes: